Dawn of the New Everything

A JOURNEY

THROUGH

VIRTUAL REALITY

JARON LANIER

THE BODLEY HEAD
LONDON

1 3 5 7 9 10 8 6 4 2

The Bodley Head, an imprint of Vintage,
20 Vauxhall Bridge Road,
London SW1V 2SA

The Bodley Head is part of the Penguin Random House group of companies
whose addresses can be found at global.penguinrandomhouse.com.

Penguin
Random House
UK

First published by The Bodley Head in 2017

www.vintage-books.co.uk

A CIP catalogue record for this book is available from the British Library

Hardback ISBN 9781847923523
Trade paperback ISBN 9781847923530

Printed and bound by Clays Ltd, St Ives plc

Penguin Random House is committed to a sustainable future
for our business, our readers and our planet. This book is made
from Forest Stewardship Council® certified paper.

To everyone mentioned in this book and the many more I wish I'd been able to mention: thank you for giving me my life.

Contents

Preface: Virtual Reality's Moment

It was the late 1980s, and a large envelope with a formidable DO NOT X-RAY sticker had just been dropped through the slot of the front door of a tech startup in Redwood City, California. The envelope contained a floppy disk that held the first digital model of a whole city. We had been waiting all morning. "Jaron, it's here, get to the lab!" One of the engineers rushed to snatch the envelope before anyone else could get at it, sliced it open, ran to the lab, and slid the disk into a slot in a computer.

It was time for me to enter a brand-new virtual world.

I squinted up at my hand against a perfectly clear blue sky. My gargantuan hand, soaring above downtown Seattle. It might have been a thousand feet from wrist to fingertip.

There was a bug, obviously. A hand should be about the right size to pick up an apple or a baseball, not bigger than a skyscraper. You shouldn't have to measure a hand in feet, much less thousands of them.

The city was abstract. This was in the early days of VR, so plasticine blocks stood in for most buildings, in a jumble of too-cheerful-for-Seattle colors. The fog was preternaturally uniform and milky.*

My first thought was to stop and fix the bug, but instead I took a moment to experiment. I flew down and tried to nudge a ferry on the

---

* This version of Seattle was built by Seattleites, just like the real version. They were researchers who would eventually become part of the HIT Lab of University of Washington, an early VR research department started by Tom Furness, a VR pioneer who had previously worked on military simulators.

sparkling Puget Sound. It worked! I had control. Not what I expected. That meant that I could still inhabit my hand when it was preposterously huge.

Once in a while, a bug in VR exposes a fresh way that people can connect to the world and each other. Those are the best moments. I always stop and linger when it happens, to hang on to the sensation.

After a few experiences of VR bugs, you have to ask yourself, "Who is it who is suspended in nothing, experiencing these events?" It is you, but not exactly. What is left of you when you can change virtually everything about your body and the world?

A bundle of cables connected my EyePhone, through a loop hanging from the ceiling, to a line of refrigerator-size computers that roared to keep cool. I wore a DataGlove on my hand; slick black mesh woven through with fiber optic sensors, and yet more thick cables from the wrist arcing up to ceiling rings. Blinking lights, flickering screens. The EyePhone's rubber rings left moist red indentations around my eyes.

I wondered at the strangeness of the world I found myself in, now back in the lab. Buildings in Silicon Valley used to have carpeted walls and cheap Space Age desks with fake wood grain. A faint smell of aluminum and dirty water.

A gang of eccentric technical geniuses converged, impatient to try. Chuck in his wheelchair; a robust, bearded lumberjack. Tom acting all professional and analytical, even though only a few minutes earlier he'd been telling me about his crazy adventures exploring San Francisco overnight. Ann seemed to be wondering why she was yet again cast in the role of the only adult in the room.

"Did it feel like being in Seattle?"

"Kind of," I said. "It's, it's . . . marvelous." Everyone shoved toward the gear. Every little iteration of our project got better. "There's a bug. The avatar hand is huge—by magnitudes."

I never got tired of the simple act of using my hand inside VR. When you could bring your body in there, you were not just an observer, but a native. But every tiny detail of functionality, of figuring out how a virtual hand could hold virtual things, turned out to be a struggle.

Fix a problem with how virtual fingertips mistakenly penetrate objects they are trying to pick up, and you might accidentally make the hand

The author as he appeared in the late 1980s outside and inside VR.

gargantuan. Everything connects with everything. Every tweak of the rules of a new world is a potential setting for a startling, surrealistic bug.

Bugs were the dreams within virtual reality. They transformed you.

A moment with a giant hand changed not only how virtual reality felt to me, but how physical reality felt. My friends in the room now looked like pulsing beings, translucent. Their transparent eyes were filled with meaning. This was not hallucination, but improved perception.

Physicality revealed in fresh light.

Introduction

What Is It?

VR is those big headsets that make people look ridiculous from the outside; those who wear them radiate startled delight at what they're experiencing from the inside. It's one of the dominant clichés of science fiction. It's where war veterans overcome PTSD. The very thought of VR is the fuel for millions of late night reveries about consciousness and reality. It's one of the only ways, for the moment, to raise billions of dollars fleetly in Silicon Valley without *necessarily* promising to spy on everybody.

VR is one of the scientific, philosophical, and technological frontiers of our era. It is a means for creating comprehensive illusions that you're in a different place, perhaps a fantastical, alien environment, perhaps with a body that is far from human. And yet it's also the farthest-reaching apparatus for researching what a human being *is* in the terms of cognition and perception.

Never has a medium been so potent for beauty and so vulnerable to creepiness. Virtual reality will test us. It will amplify our character more than other media ever have.

Virtual reality is all these things and more.

My friends and I founded the first VR startup, VPL Research, Inc., in 1984. This book tells our story, and explores what VR might mean to the human future.

The first virtual reality system, according to the original definition, in which multiple people cohabited a virtual world at the same time. This was VPL's RB2, or "Reality Built for Two." In the screens behind each person, you can see how they see each other as avatars. This photo is from a trade show in the late 1980s.

Recent VR enthusiasts might exclaim, "1984, no way!" But it's true.

You might have heard that VR failed for decades, but that was true only for the attempts to bring out a low-cost, blockbuster popular entertainment version. Just about every vehicle you've occupied in the last two decades, whether it rolls, floats, or flies, was prototyped in VR. VR for surgical training has become so widespread that concerns have been expressed that it's overused. (No one would suggest that it shouldn't be used at all; it's been a success!)

What Can a Book Do That VR Can't, at Least as Yet?

The romantic ideal of virtual reality thrives as ever. VR the ideal, as opposed to the real, technology weds the nerdy thing with the hippie mystic thing; it's high-tech and like a dream or an elixir of unbounded experience all at the same time.

I wish I could *fully* convey what it was like in the early days. There was a feeling of opening up a new plane of experience. Inhabiting the first immersive avatars, seeing others as avatars, experiencing one's body for the first time as a nonrealistic avatar; these things transfixed us. Everything else in the tech world was dull in comparison.

I cannot use VR to share what that experience was like with you, at least not yet. VR, for all it can do, is not yet a medium of internal states. There is less and less need for me to make this point as VR becomes more familiar, but it's a clarification that I have been called upon to give many times.

There's occasional talk about VR as if it is on the verge of evolving into telepathic conjuring of arbitrary reality along with a conjoining of brains. It can be difficult to explain that VR is wonderful for what it is, precisely because it isn't really everything.

Eventually a new culture, a massive tradition of clichés and tricks of the VR trade, might arise, and that culture might allow me to convey to you how early VR felt, using VR-borne technique. I have spent many hours daydreaming about what a mature culture of expression would be like in VR. A cross between cinema, jazz, and programming, I used to say.

First VR Definition: A twenty-first-century art form
that will weave together the three great twentieth-
century arts: cinema, jazz, and programming.*

Even though no one knows how expressive VR might eventually become, there is always that little core of thrill in the idea of VR. Arbitrary experience, shared with other people, conversationally, under our control. An approach to a holistic form of expression. Shared lucid dreaming. A way out of the dull persistence of physicality. This thing we seek, it's a way of being that isn't tied just to our given circumstances in this world.

If I tried to tell the story of VR dispassionately, I'd be lying. What makes VR worthwhile to me is that it's about people. I can only tell you what VR means to me by telling my story.

How to Read This Book

Most of the chapters tell a story that begins in the midsixties, when I was a boy, and ends in 1992, when I left VPL.

* This is the first of dozens of numbered definitions of VR dispersed in this book.

There are also chapters interspersed throughout that explain or comment on aspects of VR, such as a chapter on VR headsets. These "about" chapters include a dusting of basic introductory material, a hearty portion of sharp opinions, and more than a few out-of-sequence anecdotes. You have my permission to skip through them if you prefer storytelling to science or commentary. Or, if you don't like storytelling and just want to read my thoughts on VR tech, then race right to those chapters.

Some of my stories and observations are found in long footnotes. I bet you'll be glad if you find the time to read them, but you can leave that for later. There are also three appendices that expand on my ideas from the period, but are ultimately more concerned with the future than the past. Read them if you want to know what it feels like to have an informed worldview that doesn't include AI destroying humanity any minute.

In keeping with the time period of the narrative, I'll talk more about classical VR than mixed reality,* even though that's what I've worked on more lately. (Mixed reality means the real world is not hidden entirely by the virtual one; you see virtual stuff placed within the real world, as experienced lately in a HoloLens.)

---

* An example of my 1980s usage of the term "mixed reality" is found in "Virtual Reality: An Interview with Jaron Lanier" (Kevin Kelly, Adam Heilbrun, and Barbara Stacks, *Whole Earth Review*. Fall 1989, no. 64, p. 108[12]).

Meeting My Younger Self

Never thought I'd see you again.

> *What I always feared. You get old, then you milk your younger self. Like all the other writers.*

You are so wrong. It would be easier not to deal with you. I've been feeling more comfortable with myself than ever before. Dealing with you brings up crummy old patterns. I get insecure and depressed. You're recidivism bait. I'm only doing this because I think it would be useful for other people to know about you.

> *What's going on with virtual reality? Is it even called VR?*

Yeah, most people call it VR now.

> *You mean we won the terminology war?*

No one remembers or cares about that war. It's just words.

> *But is VR any good?*

Well, we're about to find out. It looks like this book might come out at about the same time that VR gets commonplace.

> *Oh crap, I hope they don't screw it up.*

Yeah, who knows . . . You know how hard it is to do VR well.

> *I hope VR isn't still so—what's the word?—pressured by all the psychedelic people.*

Oh, you'd miss them. You won't believe it, but singularity freaks cross-bred with libertarians, and their fanatical offspring are the main drivers of tech culture these days.

> *Wow, that sucks—worse than I imagined.*

I feel embarrassed that you were expecting a perfect world.

> *I'm embarrassed that you think you're noble or enlightened just because you learned to accept living with bullshit.*

Oh, c'mon, let's not fight. There are plenty of people out there to fight with.

> *Okay, so tell me about this cheap VR you say is shipping. Are people making up their own VR worlds?*

Well, usually not while they're inside, but yeah, a lot of people will probably be able to make worlds.

> *But if you can't improvise the world from inside, what's the point? Just more phenomena to clog the senses, and not even as good as in the natural world. Why does anyone care? You've got to do something to stop it before they bring out crap. What's wrong with you?*

Hey man, I'm not the VR police. I don't run the show.

> *Why not? You were supposed to run the show!*

It's actually great to watch the kids reinvent VR. There are all these cute VR startups and teams in the big companies. Some of them even remind me of you and VPL, though the fashion these days is a lot straighter.

> *I'm insulted that you'd say someone reminds you of me if that person just thinks of VR as a spectacle. Don't they know that'll turn into a cliché pretty fast? What happened to the dream of improvising reality? Shared lucid dreaming? I mean, what's the point of just making a flashier type of movie or video game?*

Look, you can't devote yourself to serving people if you think you're better than them. VR will be kind of crummy but also kind of great and it will evolve and hopefully get really great. You have to relax about it. Enjoy the process. Respect the people.

> *What a load of crap. Are you at least screaming your head off about it?*

Well, yeah, I guess . . . this book . . .

> *Okay, so who's bringing out cheap VR? VPL?*

No, VPL is long gone. Microsoft brought out a self-contained mixed reality headset—doesn't need a base station—goes anywhere. You'd be really impressed.

Microsoft? Oh no . . .

Um, my research post lately is in Microsoft's labs.

Are you institutionalized? Oh wait, you just said you are.

Give it a rest. Classic VR gear is also shipping; not unlike what we used to sell. One of the social media companies bought this little company called Oculus for two billion dollars.

Waaaaiiit whaaaat? Two billion for a VR company that hadn't shipped yet? Wow, the future sounds like paradise. And what's a social media company?

Oh, that's a corporation people use to communicate with each other and keep personal remembrances, and there are algorithms that model the people so offers can be targeted; these companies can make people sadder or more likely to vote by tweaking the algorithms. They're the center of a lot of people's lives.

But, but, combining that with VR would be like a Philip K. Dick novel. Oh my, the future sounds like hell.

It's both paradise and hell.

But bright, rebellious young people wouldn't want to be running their lives through a corporation's computer . . .

Weirdly, the new generation gap is—supposedly—that young people are more comfortable with corporations running digital society.

You say that like it's just another fact you can live with. I mean, wouldn't they become like serfs? Do they just live with their parents more, or what? The world's gone mad. Everything's inverted.

But that's normal for the world. It's what happens with time.

I feel like I need to slap you.

Maybe you do.

1. 1960s: Terrors of Eden

Border

My parents fled the big city right after I was born. They roamed for a while; eventually alighting in what was, at the time, an obscure and harsh place. The westernmost corner of Texas, outside El Paso, at the juncture of New Mexico and Mexico proper, was an outback, barely part of America. It was impoverished, relatively lawless, and of unsurpassed irrelevance to the rest of the country.

Why there? I never got a clear answer, but my parents were probably running. My Viennese mother had survived a concentration camp and my father's family had been mostly wiped out in Ukrainian pogroms. I do remember hearing that we had to live as obscurely as possible, but it would be unacceptable to live too far from a good university. They came to rest in a place that split the difference, for there was a good university nearby in New Mexico.

I remember my mother saying that the Mexican schools were more like those in Europe, with a more advanced curriculum than was available in rural Texas at the time. Mexican kids were a couple of years ahead in math.

"But Europe wanted to kill all of us. What's good about Europe?" She replied that there were beautiful things everywhere, even in Europe, and you have to learn how to not get shut down completely by the evil of the world. Besides, Mexico most definitely wasn't Europe.

So I crossed the international border every morning to go to a Montessori school in Ciudad Juárez, Mexico. Sounds strange today, since the border has come to look like the world's most advertised prison, but back then it was understated and relaxed; creaky little school buses crossed all the time.

My school was a world apart from the one I would have attended in Texas. Our schoolbooks were sheathed in fantastic images of Aztec mythology. Teachers dressed up for holidays; colorful fabrics, mod 1960s cuts, with large, living, iridescent beetles soldered to silver chains, free to wander on shoulders. Every hour or so, the beetles were offered brightly colored sugar water from eyedroppers.

Since it was a Montessori school, we were free to roam like beetles, and I made a discovery. Looking through a ragged old art book on a low shelf in our forlorn schoolhouse, I saw an image of Hieronymus Bosch's triptych *The Garden of Earthly Delights*.

Window

I remember being scolded at my little school for not paying attention. Instead I stared out the window endlessly, as if hypnotized. But this was not spacing out. It was an intense contemplation.

¡Atención! *Pay attention!*

The Garden of Earthly Delights had left me thunderstruck. I imagined being inside it, petting the soft, giant birds of intimate velvet colors, crawling through playgrounds made of transparent fleshy spheres, plucking and blowing the mammoth musical instruments that pierced each other and would eventually pierce me. I imagined how that would feel. An intense tickling, a spreading warmth.

A few of Bosch's figures look out from the canvas. What if I were one of them? When I was staring out that window, I was looking out at our supposedly normal world from inside the painting. No small chore; it took hours, infuriating the teachers.

¿Qué es lo que estás mirando? "What *have* you been staring at?"

I saw the occasional naked child alight on the little sandbox, then prance until caught, rather like in the painting. But I also saw beyond the yellow grass of the schoolyard, through the chain-link fence, to a dusty and chaotic city street.

Grizzled men in frayed straw hats inside the glass heads of mammoth trucks painted in carnival colors, blinding speeds, black noisy exhaust clouds; weathered pastel neighborhoods vanishing up into tortured striations of rock in the far desert mountainside; silver planes in the sky filled with people. Right across the street was a heroic two-story mural of Quetzalcoatl climbing a parking lot wall.

Estoy viendo maravillas. I see miracles.

Right up close, just behind the fence, I could make out more detail: furled growths on the chest of a beggar; the tottering motion of a polio survivor delivering stacks of fresh newspapers; dirt on the fringes of a teenage boy's green shirt; a pyramid of shiny green cut cactus on the handlebars of his wobbling bike. I once saw gashes in the face of a sullen prisoner in the smoky rear chamber of a careening Mexican police car, viewable for the barest instant through blinding, sweeping beacons.

Trifecta

Was everyone else in my little school blind and deaf? Why were they so inert? Why wasn't everyone else thunderstruck? I did not understand them.

I became obsessed with useless speculations. What if I had gone to a school across the river, in Texas? Things must be more orderly in Texas. If you took a copy of *The Garden* to Texas and the little naked people looked out, would they see a world that looked weird, or would they say, "Wow, we didn't know anyplace could be that boring!"

Was it possible that every place in the whole universe was wondrous, but people just get worn out by the chore of perception? Is *that* why all the other kids just sat there, pretending that everything was normal?

Of course I couldn't have articulated these words. I was tiny.

I stared and stared at the painting and then out the window and then back. I felt my interior color shift each time, like blood rushing in and out of my head. Why was the painting so luscious? What was so naughty about it that drew me in?

Even better was staring at the image while listening to Bach. The schoolroom had a battle-worn record player. One LP had Bach's organ music played by E. Power Biggs and another offered Glenn Gould on piano.

My favorite thing was staring into *The Garden* while listening to the Toccata and Fugue in D minor, loud, and eating from a bowl of Mexican chocolates, tinged with cinnamon. Hardly ever allowed.

Mood

My earliest memories are of being consumed by an overpowering subjectivity. Everything was distinct, moody, filled with flavor; each little place and every moment was a fresh spice in an endless spice cabinet, a new word in an endless dictionary.

It continues to surprise me how difficult it can be to convey a state of mind to those who do not immediately recognize it. Imagine you are hiking in the light of the full moon at midnight on a high ridge in New Mexico, looking down on a valley dusted in new snow that appears to fluoresce. Now imagine an exchange between two fellow travelers, one a romantic and the other possessing a dry, analytical temperament. The romantic might say, "Isn't this magical?" while his opposite might say, "Well, the visibility is unusually good and the moon is full."

In my childhood I was hyperromantic, unable even to conceive of a pragmatic notion like "visibility," because the experience of "magic" was completely overwhelming, to the near exclusion of everything else. My early experience was of the dominance of flavor over form, of qualia over explanation.

Over time I've learned to become more normal, or more boring. It used to be that I could hardly stand to fly from one place to another because the shift in mood and quality would be so overwhelming. I would always be stunned by what it felt like to land in San Francisco, coming from New York, even after doing it hundreds of times. The air was brisk, tinged with the smell of gasoline, but also the ocean; it was thinner, less pregnant. Just to take in the shift in feeling could take hours.

I worked at being able to suppress the overwhelming burden of subjective mood for many years, and started to make progress in my late thirties. These days, I fly from one place to another without difficulty. The airports are all finally starting to feel similar.

Spun

I called my parents by their first names. Lilly had been a prodigy concert pianist as a girl, born to a successful Jewish family in Vienna. Her father was a professor and a rabbi; an associate of Martin Buber's. They lived in a nice house, had comfortable lives. My grandparents were determined to wait out the threatening politics of their day. They were convinced there was a limit to how low people would sink.

Lilly was a precocious and resourceful teenager, and while it would normally be the last thing you'd care about, it turned out to be crucial that she was very light-skinned and blond. She was able to talk her way out of a pop-up concentration camp by passing as Aryan, and then to forge paperwork to get her father released just before he would have been murdered.

Maneuvers like this were only possible in the earliest days of the Holocaust, before genocidal procedures were optimized. In the end, most of my mother's family was murdered by the Nazis.

Some got out, eventually to New York City. At first, Lilly made a living as a seamstress; soon she had her own lingerie brand. She studied painting, and was still young enough to train as a dancer. She earned her own money to pursue these dreams. In photographs, she looks like a movie star.

We were so close that I had barely ever perceived her as a separate person. I remember playing Beethoven sonatas on the piano for her and her friends, and it felt as if we were playing them together from the same body. The interpretation was languorous and showy.

My parents had just transferred me to a Texas public elementary school. No art books to peruse, nothing interesting out the window. They were worried that I wasn't learning what I would need in order to integrate into America.

Boy, was that true. In order to get to my new school, I had to walk through the territories of neighborhood bullies. These were kids with cowboy drawls and dirty boots. I was shocked when my parents decided a karate class would be prudent for me.

I loathed every little thing about karate, except that the costumes were kind of cool. When my mother came to the faux Texas dojo to see a demonstration of my training, I stood still and took it while another boy hit, kicked, and chopped at me. I don't remember feeling scared or shy, but instead that fighting this other person would be stupid, wrong; just bad. Besides, the kid couldn't really fight and nothing he did actually hurt. But my mother was horrified; she looked disappointed in me for the first time ever. I remember feeling the sky drop.

The next morning, as I walked across the hardened soil and stubby yellow grass in our yard on my way to school, I was surrounded by big bellicose bullies. I had a baritone horn with me, which is like a mini-tuba. But to a nine-year-old, it's about as big as a tuba, and a strategy formed in my mind.

I started turning like a helicopter, the horn extended like a shield, though it acted like a battering ram. The bullies were not clear on the concept of momentum, and tried charging me head-on a couple of times, only to be knocked sideways to the ground. They couldn't find it within themselves to stop for a moment to reconsider their approach. I think there were three of them, soon bruised and running away. I was dizzy, but music had saved me.

Suddenly my self-satisfaction was atomized by shrieking. Lilly was standing behind the front door, cracked open just a bit, wailing as if the Nazis had come for me. She was not dressed and did not come outside. It took me years to realize that she must have experienced a flashback to Vienna.

At the time, I was terrified by her reaction. My not fighting at the karate studio had displeased her, and yet here I fought, but that also freaked her out. Suddenly I felt a disconnection. The sensation was so disorienting and unpleasant to me that I didn't know what to do. I ran away, to school. That was the last time I saw her.

Beyond Recall

A glum man with sharp features, in a perfectly pressed military uniform, knocked at the door of the classroom and asked for me. I was happy to get out of a droning lecture about the Alamo, but something felt terribly wrong.

Soon I saw that the principal was also there, and this man asked me in the most formal voice I had ever heard to follow them to her office, where I had never been. There was a flag, a framed photo of President Johnson. Was I in trouble for hitting the bullies with the horn?

Then these strangers told me that my mother was dead and my father was in the hospital.

It happened to be the day Lilly was set to go into town for her first-ever driver's test. The DMV was about an hour away, near downtown El Paso. Ellery, my father, drove on the way there. She passed the test.

Lilly was driving back, on the big freeway, when their car spun out of control, flipped, and flew off a high overpass. Or so said a fresh newspaper clipping, which the principal gave me, as if that was helpful.

For years I worried that Lilly's traumatic flashback that morning might have made her panic on the road. I was consumed by guilt. Had I been part of the problem?

Decades later, an engineer friend of mine read about a possible flaw in the model year of the car. It was a match for the events of the accident. By that time it was way too late to look into legal recourse, but I wondered, why did my parents even buy a car from Volkswagen? It wasn't a "beetle," the model designed by Hitler, but still.

The choice must have been part of my mother's program to find the good in Europe, in everything.

It turned out that the military man was a distant relative the police had tracked down. He was named in my mother's will, and was stationed at Fort Bliss, the military base that makes up much of El Paso. I had never heard of him.

I was taken to the hospital to see my father, whose body was blackened in between bandages, after he had become conscious. Both of us cried uncontrollably, so hard it felt like I'd choke to death.

This memory is a wall. I remember almost nothing else from before my mother's death. My slate was shaken clean.

Sound

I was disconnected from the world for a long time after. Endured a desolate tour of life-threatening infectious diseases, barely aware of my circumstances. I was virtually immobile for a year in that same hospital.

Ellery was devoted. He slept on a cot next to my hospital bed. The seasons cycled, and I finally started to engage with the world again. I remember paying attention to my new surroundings for the first time.

The hospital was cramped, hot, and noisy. Cracked pea-green tiles running halfway up the wall, greasy windows embedded with chicken wire, splintered frames, peeling dark green paint. Smelled like medicine and urine. Big tough nurses with tiny crosses hanging from their crinkled necks; they moved like tanks, mostly ignored everyone.

I started to read. Books propped on crimped-up bedsheets.

Then, two moments of irreversible positivity sparked in me, just because I read sequences of words.

One was the Jewish admonition to "choose life" in a children's book about Jewish culture. There was a logic to it, since death would come soon enough, no matter what, so choosing life was at least a reasonable bet. Like Pascal's Wager, but for this life. (Not that I would have heard of Pascal or his wager as a kid.) But as I thought about it, I realized that "choose life" has even more going on.

It's so obvious that you could miss it, but the phrase tells you that life is a choice. Furthermore, it suggests that once you notice that you've chosen to live, then you might notice that you can probably also make further choices. I needed to hear that, because it had not even occurred to me that I had any choice at all at the time. Before reading those words, all I could do was lie there, waiting for whatever might happen next.

But then there's an even deeper level to the phrase. You choose even though you can't ever know what that means. This physical world we inhabit; we're only in it because of a crazy bet we make with the unknown. Maybe there's peace and happiness to be found in uncertainty. There isn't anywhere else to look.

I guess you, my reader, might wonder if I'm forcing adult thoughts into my recollections of a child's brain, but I remember this phase pretty clearly. I was obsessed with what's usually called philosophy, and it helped.

The second bit of reading was a biography of Sidney Bechet, one of the great early New Orleans wind players. According to the book, he overcame his childhood respiratory problems by playing the clarinet. Well, I had a nasty case of pneumonia that persisted for months, along with other respiratory distresses, so I asked Ellery for a clarinet. Not only was it a great way to annoy the nurses, but my lungs started to clear.

This is starting to sound like a familiar inspirational healing narrative, but there is something else you should know. My father and I never again spoke of my mother.

Silence is not forgetting in intimate settings. Just the opposite. We still lit Yahrzeit candles; we cried for years.

Decades later, I realized that both my parents had no choice but to put those who died out of mind much of the time. It was the only way to make space for life, for there were so many who had died so horribly.

Ellery had an aunt who was entirely mute, but she wasn't born that way. As a girl she had survived by keeping absolutely silent while her big sister, who she clung to, was slain by sword where they hid under a bed during a pogrom.

To Ellery, Lilly's death was one of many. Therapists, daytime TV hosts, and social networks all counsel that we should talk. It's a luxury when you can afford to do it.

Outlasting Cruelty

When I emerged after my long dormancy, hosting one disease after another, I was fat, but didn't know it at all. I was numb. When I finally went back to school and the kids laughed so cruelly, only then was I suddenly aware.

Normally, taunts from children would be traumatic, but there was more. Juvenile cowboy bullies bragged about how they drowned a tiny Chicano kid in the neighborhood swimming pool, an event that the adult world had officially registered as an accident, though everyone knew.

The bullies said I was next. They were credible. There were only a few Chicano kids, and they'd come in with casts and scars, and look away from everyone.

One of the teachers pointedly reminded us in class that the Jews killed Jesus and were still paying the price. She then said this ancient, cosmic crime probably had something to do with my mother's accident. My mother had had it coming.

I realize now that this teacher was making her best attempt at kindness. It was her way of saying I couldn't help that I was born Jewish. In a similar vein, she urged the white children to be understanding, since Mexicans couldn't help the fact that they were less intelligent.

Thereafter, I was bombarded with demands that I convert. My memories of that school are of constant onslaught, racism, and violence; of adults who were no better than the children.

I was a few years younger and therefore smaller than the other kids in my grade, and an easy mark. A notorious cowboy kid confronted me, with a crowd urging him on. He was a dandy, in a kid's black western shirt. I suddenly recalled lessons from the karate school, which by then had taken place so long ago, in the unreal antebellum. I built up tension in my whole body in order to aim a punch and knocked the guy flat on his back.

It would be nice to tell the Hollywood story here. Wouldn't I suddenly be an alpha, raised on shoulders, loved? No, I found myself more isolated than ever. I was regularly ambushed and beaten.

The thought of connecting with other people—of having friends—was terrifying, and strangers were dangerous. It was impossible to say how much of my dread was due to my circumstances and how much was inherited from my parents.

Reality is ever changing. The odd demographic churnings of the region eventually brought me into contact with a variety of people, and I slowly learned to connect pleasantly with oddballs I stumbled into. One time, I wandered into the Radio Shack store on the main road and met a polite soldier from Fort Bliss in a not-so-fresh beige uniform.

The guy was awkward; looked down all the time and walked as if the floor was being tilted back and forth just a little bit by a supernatural prankster. He noticed me yearning at the drawers of electronic parts and said hello.

He seemed young, even to me. Not quite a real mustache yet. Worked with radar equipment. Wouldn't tell me more.

What makes people generous? What makes a stranger take a chance? This fellow took it upon himself to introduce me to electronics. He brought a few parts to our little house a couple of days later; resistors, capacitors, wire, solder, transistors, potentiometers, a battery, a little speaker. We made a radio.

First Experiment

Right next door to that Radio Shack was a drugstore with a magazine stand. Before the Internet, magazine stands were a big deal. You could go and browse the covers without even touching; show dogs, fancy boats.

The rack itself was made of thick, curled, shiny wire; fancy but cheap-looking at the same time. You had to be choosy about when to go, because the afternoon desert sun blasted through the big window. It would glint off the wire, making it painful to look.

There was always more to learn about the magazine rack. The magazine covers lost their warm inks and turned blue after about a week in the strong sun, so you could always tell how long an issue had been out. There was supposedly a room with dirty magazines in the back, but I never saw it.

A few titles covered the amateur electronics beat. Most of the articles were about building radios, but I found a piece about an early electronic musical instrument called a Theremin and learned how to make them. Theremins are played by moving your hands in the air near antennae; nothing is touched, and playing gives you the feeling of contact with a virtual world.

I was also fascinated by the diaphanous, silky, turbulent images called Lissajous patterns, which can be made by fiddling with musical signals and an oscilloscope. I made a crude Lissajous viewer out of an old television set I found in a trash bin, and hooked it up to a Theremin. Normally, a Theremin makes spooky warbling sounds, but I got one to make spooky warbling images.

As Halloween approached, a plan formed in my mind: I would build a fantastic haunted house out of my electronic contraptions and attract people worthy of being friends! There must be others like that kind soldier, just wandering about, out of sight, like desert tortoises. All you had to do was find them.

I hung sheets around our tiny front porch and set up an old enlarger lens to project Lissajous patterns from the TV onto them.

Once the sun went down and the images appeared bright, I was deliciously surrounded by fantastic dancing forms. The motions of anyone nearby would alter the patterns, as if by the invisible strings of a puppeteer, courtesy of the magical Theremin antennae.

I wondered whether any girls, those beings of utter mystery, might be delighted. Who wouldn't be?

My haunted house pleased me immensely but attracted no visitors. I watched from inside my palace of imagination and freedom as one child after another crossed the street to stay as far away as possible. It never occurred to me at the time that they were probably frightened. They had certainly seen nothing like it before.

After that Halloween, the bullies stopped bothering me. I had made myself into a scary unknown. Progress.

Burned

Another astonishing thing about my mother was that she was the bread-winner of the family, at least when we had moved out west. In that era it was *always* the man.

There was a truckload of ill will directed toward Ellery over this matter, before her death and after. "The boy needs to see his father earning an honest living, being strong. You're letting him down. Keep this up and he'll grow up *funny*." Good citizens were so sure of themselves that they didn't care if I was right there, listening to it all.

My mother made our money over the phone, trading stocks in New York, back when no one did that. She wasn't a tycoon, not even close; we were middle class, but not in the upper reaches of the middle. We could eat at a drive-up hamburger joint every week.

Players are usually either well-off or wannabes on Wall Street, or any other wide-open platform, but my mother found a do-well-enough niche. Could she have done better? Maybe she was afraid of standing out, being seen.

I do remember a little more than I've said so far. I remember her getting off the phone one day and exclaiming that she'd just closed a great trade, made not just hundreds, but a few *thousand* dollars. That call persists in my recollection because it yielded the money that bought the car she died in. We went out the very next morning to buy it. I got to pick the color.

After my mother died, there was a secondary crisis, because Ellery and I no longer had income. While I was in the hospital,

Lillian Lanier.

Ellery enrolled in a program to get certified to be an elementary school teacher. So that was a solution to the income problem, but then something else went wrong.

We had known for a while that our lease would run out and we'd have to move. This had happened a few times. My parents finally arranged to buy a house so we'd never have to move again unless we wanted to.

It was a tract house under construction in a development at the edge of El Paso, entry level, but upscale compared to where we had lived before. It had a carport! I visited it only once when it was still under construction, but I was fascinated by the blueprints and pored over them. I learned all I could about drafting and construction. I couldn't wait to move in.

While I was in the hospital this house was completed, then burned down the next day. Ellery told me but the news didn't register. I thought it must have been a dream, and was still confused about the facts when I was discharged.

The police informed Ellery that the fire was arson, but there were no witnesses and no suspects. Ellery muttered that we might have been targeted, but it could have been random. Bad things happened all the time around there.

There was a screwup with the bank or the insurance. After the incineration we didn't recover any of the money my mother invested in the property. Ellery felt especially bitter about having to pay to have the ruins cleared.

Thus it came to pass, not long after my haunted house experiment, that we had to move but had no place to go.

2. Rescue Spacecraft

Landed

Ellery did something unthinkable and brilliant. After my mother died, and our new house burned down, and we were broke, and I oscillated between terror and isolation—after all that, he bought an acre of throwaway land in New Mexico.

The lot was cheap enough that he could buy it with the small remaining amount of cash on hand. *And* he found a teaching job in the same area.

An undeveloped corner of the desert. We had no money left to dig a well, much less build a house. We had to live in tents at first. Our belongings, even my mother's baby grand piano, were wrapped up in plastic on pallets out in the open on the indifferent desert dirt.

Ellery took to teaching sixth-grade kids in the rough little barrio in the center of Las Cruces, New Mexico, as if it were an art form. He had tough kids build cardboard spaceships to inhabit all day long. They launched model rockets and used sand to explore the ideas behind calculus. He was known as *pelón* to the kids, for Ellery was shiny bald, like a polished gem.

Whenever I go back to visit Las Cruces, people come up to me, and with that distinct New Mexico Chicano accent, say things like, "Your dad Ellery changed my *life*. My older brother is in jail, but I'm a NASA engineer."

We ended up living in tents for longer than planned, for over two years.

When Ellery's earnings as a teacher started coming in, the first priorities were a shed where electricity and a phone could be connected, a well for water, and an outhouse.

The high desert can be bone cold, and I remember shivering like a spring marionette on winter mornings. The people who bought lots around our land wheeled in trailers and mobile homes. We talked about it. We could do the same, but that would mean diverting money from the *big plan*. Not worth it.

We grew vegetables. Raised chickens.

Tent life was not so bad. It clarified your role in your own survival, and we both needed that. And those rolling homes were so *ug-ly*.

Where in the Universe?

There was a social anomaly in that part of New Mexico: a population of superb engineers and scientists employed by White Sands Missile Range. They were everywhere. It was a relief to discover the culture of technical people, which was welcoming to an awkward kid like me.

One of our near neighbors was a lovely, slight old man named Clyde Tombaugh, who had discovered the planet Pluto in his youth. When I knew him, he directed research in optical sensing at White Sands.

I learned how to grind lenses and mirrors from Clyde, and I still think of him when I work on virtual reality headset optics today. He built impressive backyard telescopes, and he let me play with them. I will never forget a globular cluster he showed me—a vividly three-dimensional form, a physical object like me, a cousin to me, as real in front of me as anything else in the world. I gained a sense of belonging in the universe.*

I went to public school in New Mexico, and I don't remember a lot about it, which probably means it was okay. At the very least, I didn't experience terror.

Right after we arrived, before I got to know any other kids in the area, an amazing thing happened. One evening there was a perfect breakdown

---

* I have no sympathy for the recent campaign to demote Pluto to prominent Kuiper Belt object instead of planet. Its weird orbit out there is an inspiration to every kid who doesn't fit in. Are we not full-fledged planets? Will you only accept us if we conform? Let Pluto remain a planet, now and forever! If you planet demoters want to campaign to make folk categorizations of our world more rigorous, why don't you insist that Europe isn't a continent? That would be more useful.

of the local telephone system. Anyone who picked up a phone could hear everyone else, all at once.

Hundreds of voices—some sounding distant, some close by—hovered in the first social virtual space I had ever experienced. An instant society of children formed, brilliantly superior to any I had experienced before.

The floating children were curious about one another; they were friendly. It was less fraught to communicate with strangers than in real life. The voice of a little boy said, "I've hugged every woman in the world as a pillow." And this was with real girls floating nearby.

It was late at night, and none of us were supposed to be up, though I might have been the only one in a tiny plywood shed that could only be secured by padlock.

The next morning at school, no one spoke of what had happened. I looked around and wondered who I might have talked to the previous night. Was it possible that people could suddenly improve, if the medium that connected us was different?

Ever since, I've tried and tried to rediscover that formula. Maybe it was a one-time benefit due to novelty. Whatever was going on, it's long been clear that it's easier to design a virtual space to make people worse.

A lot of folks we met near our land beheld apparitions. I'd walk home from school along an irrigation ditch, so I'd come across men taking breaks from working the land. They'd talk about the weather or cotton prices, but just as often about miracles.

"You know Alicia, she almost died in the hospital, but the *curandera*, she said Mary would come to her, and then she came, shined like the sunset, and Alicia got real better. Now she's bothering me every day. Like I don't work hard enough."

The story went on indefinitely. I'd wait for a moment when I could say 'bye, but there wasn't one. You just walked on, maybe with a slight upward sling of the head, as if your chin was pitching an invisible ball.

The border region was crawling with spiritualists of every kind; evangelical, pueblo, Catholic, hippie. That could mean trouble. I once became enraged with a shaman from the Copper Canyon region of Mexico. Had a prosthetic agate eyeball and dressed in ribbons; claimed to have made contact with my mother, wanted money. I think he might have even gotten some out of Ellery; we were both still vulnerable, went through patches when nothing made sense.

At least you could trust the sincerity of murderous kids in a school-yard. Friendly people could be sneaky. A difficult lesson.

There were also secular apparitions; a local culture of flying saucers. Kids would bring bits of fallen alien spacecraft to school for show-and-tell and no one questioned their authenticity, certainly not the teachers. We lived next to the largest missile test range in the world, so peculiar debris fell from the sky all the time. I still have beautifully machined satellite detritus I found in the mountains.

I never believed they were really alien, but I did find myself entering into the cult of local pride about *our* flying saucers. I still feel an involuntary surge of indignation whenever a rival town, Roswell, New Mexico, gets renewed attention for its inferior 1950s flying saucer crash landing. Our flying saucer crashes were better!

Whence

It must have seemed to Ellery that he had been preparing to live in New Mexico for years.

Before I was born, he had undertaken a wide range of careers simultaneously, just as I have. He studied architecture at Cooper Union and built skyscrapers with his father, who was also an architect. He also had a job designing window displays at Macy's, and he and Lilly had shown their paintings—cubist—at a few notable shows.

Ellery had a mystical bent. He had lived with Gurdjieff in Paris and Huxley in California and studied with various Hindu and Buddhist teachers.

Hand in hand with his interest in mysticism, which he distinguished from superstition, Ellery liked confronting hokum. He was a minor radio personality in the 1950s, a semiregular panelist on one of the very first radio call-in shows, hosted by the pioneering broadcaster Long John Nebel, who was known for his interest in the paranormal.

They had great fun humoring crank UFO and paranormal enthusiasts on the air, but ultimately exposed con artists. Ellery wasn't above making up nonsense as a gag, along the lines of the *War of the Worlds* radio show.*

---

* This was Orson Welles's notorious 1938 radio drama that simulated an alien invasion with enough bravado to cause panic in a gullible public.

He claimed to have made up the alligators-in-the-New-York-sewers urban myth. Could be so.

Once, he exclaimed on live radio that a noisy purported antigravity device had just maybe lifted a bit. He then had to explain it was a joke after callers took him seriously, but they couldn't be talked out of believing.*

Ellery also wrote columns for Hugo Gernsback's pulp science fiction magazines of the 1950s. He was, for a spell, the science fact editor of *Amazing*, *Fantastic*, and *Astounding* magazines. He'd explain the science relevant to the stories in each issue—for instance, the latest research on Mars when Isaac Asimov set a story there.

One of his columns was about "making your own universe." It described a recipe for a murky fluid you could mix in a big glass jug. Stir, and tiny formations resembling galaxies would form inside.

Ellery was part of the social circle of New York science fiction writers. They were pranksters; had a betting pool over who could make money in the most preposterous way.†

Asimov took a minimalist approach, taking out an ad that said, "Quick, send a dollar to this post office box." No explanation, and yet the dollars flowed in.

Ellery and Lester del Rey advertised a service to have a baby's first dirty diapers bronzed. People would send the money in advance. The diaper itself, once it was skillfully prepared by a baby, went to a different address, which turned out to belong to the American Nazi Party.

Permission

The *big plan* was precisely insane and the only conceivable path forward. Ellery would let me design a house. I'd have to submit a design and get it approved by the county. We'd gradually buy building materials as we could afford to. We'd build the structure with our own hands, however long that would take, and then move in.

---

* One day in the late 1970s I made Ellery call in to the Nebel show, which was still on the air. He, Lester del Rey, and Nebel all started hurling insults at each other, and I could understand why the show had been so popular.

† L. Ron Hubbard was an early member of the circle and played around with promoting his ideas as part of the betting pool, gaining skills he later applied with great amplification.

Ellery had studied architecture and had helped *his* father on projects like extending the height of a skyscraper in New York City. But he realized that I needed a meaty obsession if I was ever going to become fully functional again.

As a start, he gave me his copy of an old book he loved when he was a child, called *Plants as Inventors*. It included delicate illustrations of botanical forms. I was mesmerized. Some of them looked like they would have fit in well in Bosch's garden.

The spherical designs were especially enchanting. There are only five ways to divide a sphere with perfect regularity. This had been known since ancient times, and the flat-faced versions of the five solutions are known as the Platonic Solids. Plants had no choice but to work within the constraints of these forms.

I became convinced our home should be made of spherical structures resembling those found in plants. Ellery said he thought I might enjoy another book, in that case.

This turned out to be a roughly designed publication in the form of an extra-thick magazine called *Domebook*. It was an offshoot of Stewart Brand's *Whole Earth Catalog*.* Buckminster Fuller had been promoting geodesic domes as ideal structures, and they embodied the techie utopian spirit of the times.

Initially I was skeptical of going geodesic. "I don't want our house to be like any other house, and other people are building geodesic domes," I complained. Ellery argued that I'd have to get approval from the authorities to build a design, and a few geodesic domes were already standing in hippie enclaves in the same county. Including this counterculture cliché would inoculate my design, make it less scary.

I started to make models out of straws, and then calculated angles, loads. I must clarify that I wasn't necessarily carrying out these calculations correctly.

My design strategy was to mix "conventional" geodesic domes with connecting elements that would be profoundly weird and irregular. There

---

* This was Stewart Brand's humongous book that you could browse for hours; filled with portrayals of people doing interesting things and interesting things you could buy from them. It suggested a comfortably ambiguous utopian principle, in which people retreated back to the land but were also futuristic. The tome is sometimes remembered as a paper prototype of the most colorful aspects of the early Google, or at least that's how Steve Jobs would later frame it.

was to be one big dome, about fifty feet across, and a medium-size one, to be connected by a strange passage, which would serve as the kitchen, formed out of two tilted, intersecting nine-sided pyramids. There would be two icosahedrons, twenty-sided forms, connected to the big dome by another set of complicated shapes. The icosahedrons would be bedrooms, and the interconnections would contain a bathroom.

A cantilevered, blade-like, seven-sided pyramid would jut out, carefully formed to point at certain astronomical objects at certain moments, but I don't remember what these were! Too much time has passed. You'd enter through a door in the side of the jutting structure, which would be called "the needle."

The overall form reminded me a little of the Starship *Enterprise*— which has two engines connected to a main body and a prominent disc jutting out in front—*if* you filled out the discs and cylinders of that design into spheres. I had always thought that that should be done anyway, since a starship moves in deep space, not in an atmosphere above a planet. I put cantilevered forms everywhere in the nondome parts, to try to create the illusion that the craft had not quite completed the process of landing.

The design was also a little like a woman's body. You could see the big dome as a pregnant belly and the two icosahedrons as breasts.

At any rate it was a form that I liked and that Ellery accepted. There were a few passes back and forth with the building permit people, and ultimately Ellery did have to intervene to argue the case, but we got a permit.

One of my straw models, lying on the ground where the real version would be built.

Build

I wish Ellery hadn't given me that *Domebook*. It pretended to offer solutions when it was actually reporting on ongoing experiments. The *Domebook* advocated ferrocement, a boatbuilding material. I should have learned about that material from a shipbuilding expert, but instead trusted what the book said.

"Hog rings are the thing!" Thus spoke the *Domebook*. These are the little rings you crimp on a pig's nose to make it less destructive. The innovative idea was to fasten sheets of metal lath together with hog rings and then press concrete into the layers. Bad idea. The lath density was inconsistent and the result was subject to cracking.

About ten years later I'd meet Stewart Brand for the first time, and my first words to him were, "I grew up in a geodesic dome." His first words to me were, "Did it leak?"

"Of *course* it leaked!"

We started first on the medium-size dome, because that was all we could afford. It was a strange feeling to move into it from the tents, like recapitulating deep human history.

The inside of the dome was insulated with shiny silver-surfaced pads that were stapled between struts. These were meant to be covered by sheetrock, but the money and trouble to sheetrock the inside was beyond consideration. So the interior was baggy, shiny silver, like a space station. Perfect.

After another year it was possible to buy materials for the rest of the house. I remember pouring the cement foundation for the big dome, and desperately trying to get all the little desert springtime frogs out in time so that they wouldn't be entombed. We had to climb up on the strange, unfolding, triangulated framing to raise the structure, and it felt like being a spider mastering a web. Neighbors talked about *arañas en el cielo*, spiders in the sky.

We mounted protruding, hemispherical windows, a phantasmagoria of odd touches.

Habitat

The place was huge for a residence. The larger dome was big enough that you could almost focus at infinity while staring up at the curve of the

The author, age approximately thirteen.

baggy silver ceiling. It created an illusion of a solid sky, a little like being in the Big Room at Carlsbad Caverns.*

We called it "the dome," or "Earth Station Lanier." One would "go dome" instead of going home.

The dome was a cabinet of curiosities that had been tossed in a blender. An old telescope that was supposed to have been the one used by Commodore Perry to first sight Japan was there, given to Ellery when he wrote an article about Perry's family. Maybe it's authentic. I still have it. I screwed it up a bit when I tried to mount it at age twelve or so.

There was a scrap that supposedly came from an original Hieronymus Bosch canvas, and wonderful old antiques from Vienna. A decade after the war, a Good Samaritan came across a few of my grandparents' items that had been seized by the Nazis and offered to ship them to my parents in New York. An ornate alarm clock, a fancy chest. There were also vast, colorful geometric models, biofeedback machines, many, many paintings, Ellery's old experimental color organ, and mountains of books.

There wasn't a proper bathroom or kitchen. Instead, tubs, sinks, and showers were inserted into the structure according to how the plumbing

---

* The giant cave of dreams for every kid growing up in New Mexico. So big that the sky is rock. A friend from Italy said it was better than the Vatican.

Ellery in front of the nearly completed earth station.

Inside the big dome.

could be routed through the bizarre shapes I had chosen. A sink was unusually high off the ground; you needed a stepping stool to use it. Conventional choices regarding privacy, sleep schedules, or studying were not really possible.

I loved the place; dreamed about it while sleeping inside it.

Years later I would realize what a leap of faith Ellery had taken to let me design our house. He could have intervened more, but I think he wanted me to learn to take risks and make mistakes.

If that was the idea, he succeeded a little too well. When I moved away, Ellery decided to continue to live in the structure. He stayed there for thirty years before parts started to fail. Once, he had just walked outside when the first ring of the big dome, closest to the ground, collapsed with a roar. The whole half sphere just settled down as a unit, losing vertical height but not damaging anything inside, as if it were a cartoon house on a pneumatic lift. By the time I got there to look, he had already replaced it with a new dome.

The needle had been lost, however, as well as some of my other bizarre shapes.

Ellery continued to teach as long as he was able, for he had found his vocation. After he retired from the New Mexico public schools, he taught elementary school on the White Sands Missile Range. He stayed in the dome into his late eighties, when he couldn't live independently anymore.

As for me, I moved away, but never fully. Having grown up in such an odd environment, I found it quite a challenge to live in a normal place. I had a hard time adjusting to orthogonal walls, and normal schedules. I spent much of my thirties forcing myself to live more conventionally, without clutter. Then I met my wife, who has a mother who is a neat freak, so in order to compensate, she enjoys clutter. We've extended our home to include a structure that is reminiscent of the needle. We live back in the dome, more or less.

3. Batch Process

From Atoms to Bits and Back Again

I had barely turned fourteen when I attended a summer camp for chemistry at the local university, New Mexico State. There were hundreds of kids from all over the country. Well, given how memory works, maybe there were only dozens.

We were moved around in buses a lot. I looked out through rows of slanted windows, lined in shoddy chrome, to see sand and cactus swirling subtly in the distance as we made our way up mountain roads. I imagined being a photon, my path perturbed by desert thermals.

I was used to the scenery, but was fascinated by how streaks of fierce sunlight scanned shapes inside the bus. Sunlit faces of children became translucent; thin slices of animated agate.

As we were jerked about on dirt roads, the exhaust mixed with sage. A visit to the telescopes on top of the turtle-shaped mountain, to the actual white sand, then to the missile range named for the sand. I was the only local kid, so was able to feel a little more knowing than the rest, for once.

I remember meeting a pair of exquisite freckled twin girls from Colorado, and they talked to me as if I were a normal human, even though I was a few years younger. "Both our parents are chemists!" This was terribly strange, and a pleasure.

Chemistry was pure beauty and intrigue for me. It turned out that our universe's particular batch of elementary particles happened to be able to form into interesting atoms. They did so by creating fantastical shapes, the electron shells. Those atoms happened to be able to form themselves into interesting molecules that could evolve into us.

My father and I had just built an elaborate, barely functional structure out of the same crystal symmetries found at the core of nature, so I had a keen sense of how easily such designs can fail. The whole scheme of reality seemed so unlikely. The particles themselves didn't have a chance to evolve, so how could they be so perfectly set up for this big show of ours? All it would take is one little change and the whole universe would crash, just like one false bit can crash a program, or one hog ring can crash a geodesic dome.

There's always an answer for a question like this. Many years later I'd meet a physicist named Lee Smolin who proposed that universes could indeed evolve, by birthing new universes inside black holes, to settle upon sets of particles with interesting properties.

I was in a constant state of awe. I learned to make various chemicals, the usual things like fruit odors and explosives. "Mr. Lanier, would you please consider concluding your experiment today in the vacant lot across the street?"

At the end of the summer it was unthinkable that I'd return to high school. I just stayed in college.

No acknowledgment of high school equivalency, and indeed no process of admissions took place. I fudged my way into signing up for classes. I don't quite remember how it happened. Maybe I was supposed to be in high school while I took college courses, but I signed up for enough courses to be full time, and never went to high school.

Through whatever fluke or forgery, long forgotten, I quickly became a full-time college student.

Access

There were wonders to explore.

There was a music department where I could take composition courses. I studied species counterpoint and orchestration. I was interested for a while in writing tiny piano pieces, like Satie or Webern for a mouse. A composition teacher repeatedly demanded I make them longer, which I did, over and over, until one day he said, "Mr. Lanier, I'm surprised at you. This piece wanders."

There was a locked room where the less-used orchestral instruments waited for their occasional Cinderella moments. I gained access and practiced on contrabassoon, celesta, and the other wonderful musical machines we inherited from high European culture.

The clarinet had probably saved me after my mother died, but she also left me a folky Viennese zither painted in flowers, a violin, and a piano. The piano I played with complete devotion and seriousness, though after my mother's death I found that I couldn't resume the path to becoming a classical pianist. Instead I veered into bizarre, furious improvisations.

The zither I treated as an experimental instrument, hammering on it with the back of the handle of the tuning wrench, producing a heroic-sounding effect that I thought would make a good soundtrack for a Superman movie. There was only time for a preliminary violin lesson or two before she died, and I simply couldn't even look at the violin for decades, though I kept it. Now I am glad, for I am finding great joy in being able to take up the instrument as a beginner in my midfifties.

There was also an electronic music lab with a Moog modular synthesizer, among other jewels. (I noted that universities felt compelled to buy certain expensive pieces of equipment. I'd eventually have them doing that with virtual reality systems.)

Bob Moog created one of the enduring languages of technology with his simple set of synthesizer modules. I had tremendous fun with them and created curious pieces of music on tape. You could set up a feedback path such that the synthesizer would settle into an equilibrium so sensitive that just clapping in the air nearby would incite it into a convulsion.

In the math department, odd men with beards spent days struggling to prove theorems concerning Abelian groups. Being around that process in the math building, well before I could understand it, felt like gaining entry to the inner sanctum of a temple. I was where I wanted to be. I couldn't sleep all night for the joy the first time I understood why e to i times pi was minus one. Ellery had told me, but I wouldn't believe it until I could "get it" for myself.

The Nasty Bits

NMSU also had an early and fine computer science department, a result of the proximity to the missile range.

At first, computer science seemed like a lesser pursuit than mathematics or chemistry. A study of human inventions like computer programs was subordinate to the study of truths that towered above humans.

Despite that, I thought computers might address anxieties that drained me. As a fourteen-year-old I worried about Earth's orbit. So precarious. We were just spinning in space, it seemed to me, and any heavy object that happened to come near might send us plunging into the sun. It hadn't happened in billions of years; nonetheless, I wondered about devices to protect our orbit, should that be needed someday; there would have to be an automated adjustment system, and it would have to be controlled by computers—so I decided I must study computers.

At the time, the most common way for a student to use a computer was to take a deck of punched cards to a service window and hand them to a tech. The tech would eventually hand your deck to a more elite tech who would feed it to the exalted machine that mere undergraduates didn't even get to see. You'd have an appointment to pick up a result in the form of more cards.

Desert winds bite. You'd have to lean while you walked, and your windbreaker would whack back and forth like a sputtering motor. It was not unusual to see flickering cyclones of punch cards in the sky, some flying like squirrels. Panicked students would scream and run after them, but I doubt anyone could ever have reassembled a deck once it was blown. It happened to me once and I cheated by performing the program myself instead of admitting I'd lost my cards.

One day I was waiting for my turn to go up to the window, and my punch cards were stacked under a frayed copy of *Gravity's Rainbow* on the pockmarked wooden shelf that ran the length of the wall, under rodeo and football posters.

Thomas Pynchon, who wrote that novel, never appeared in public. No one knew what he looked like.

The guy behind me in line murmured, "That guy is such a dick."

Who, me?

I turned and looked up at a military man. In uniform, nerd glasses, intense eyes, perfect blond trim, obviously smart.

I managed "Um, wha?"

"Pynchon! He doesn't let us see him. Information asymmetry! He sees us, we don't see him. He's on a power trip." How could he say something like that about such an amazing writer?

"Novelists aren't powerful, right?" I said. "I mean, he probably just doesn't want to be bothered. It's not like he's got missiles."

"You really don't get it. Amazing."

I attempted one last stand. "If a writer wants to stay out of sight, isn't that harmless? It's just a little bit of a curtain, like a fig leaf added to an old sculpture. It's not like we'd see anything that would really matter."

"Fig leaves are the ultimate information weapon. You obviously don't understand anything, kid."

It was finally my turn at the punch card window.

"Um, good to meet you. What's your name?"

"You'll never know, kid."

I wonder what became of him.

Goats

Tuition was low—a simple, easy fact of life in those days that now feels like a mathematical impossibility—but I needed to pay it. Ellery didn't earn much as a teacher, and the whole college thing was my folly. The answer was goats.

I had made friends with a goat who resided near the dome. She was a cute Toggenburg, like a fawn, with a generous personality.

It was impossible not to get a goat, then another. Herds are traditionally named, and mine was registered as the "Earth Station Goat Herd."

The next step was learning to make cheese, and then figuring out how to sell it. There wasn't much competition, but there was demand. Easterners moved to the desert for health reasons, and they sometimes preferred goat dairy products. Easier to digest.

A local hippie food "co-op" was my main customer, along with individuals who would come by. I earned enough to make it work. I kept my expenses low.

It might sound strange to start a goat dairy to pay for college, but there was an agricultural strip right along the Rio Grande. NMSU also had a big agriculture school (the football team was called the "Aggies") and what I did was considered fairly normal.

Milking a herd of goats every morning and evening is quite consuming, though. Also, trimming their little hoofs, tossing around bales of hay. But I loved my goats.

You can choose not to believe me, but it's true: My goats knew their names and were housetrained. Many were named after stars in the Pleiades. Alcyone, Merope . . . I learned to make goat calls and talk to them. They were Nubians, and so had plaintive, almost alarming voices—like a wailing Armenian duduk—instead of the more common *me-he-he-he-he-he*. I would call them in turn, using both English and faux goat, and they'd rush into the dome and up onto a milking stand, where I'd milk them in cleaner circumstances—and with the milk refrigerated more quickly—than in other goat dairies.

I'd also play flutes for them, like Pan. I was proud of the Earth Station goats, but tortured by the idea of slaughtering a kid. Alas, there wasn't much other economic use for most of the males. So I researched every old wives' tale from the extensive world of folk goat husbandry. I fed them weird vinegary concoctions, and encouraged them to jump up on the dome. While my sample size was too small for publication, it worked! I had almost no male kids. Am I going a step too far by mentioning that one of my goats, Onyx, won Best Udder in the New Mexico State Fair one year?

Students were required to choose either a sports or home economics elective. There was no way I was going to engage in sports with the much older, often quite macho guys at a big ag school. So I ended up the only boy in a sewing class. I was much younger than the girls and they found me adorable. Had I been the same age, I suspect I would have been mocked. I sewed my own clothes for a while, saving money. I remember making goofy Robin Hood–like cloaks.

I started out younger than most college kids, but after a couple of years I started to fit in. The life of a young adult finally unfolded for me, though not in an entirely normal way.

Ellery taught me to drive, but that rite of passage was dominated by terrified yelling. "You have to be ready for anything to go wrong at any time. The other people on the road might turn out to be drunk, or murderers. Your car might suddenly explode."

The university was the relatively calm and safe place that I needed.

When I interview brilliant young people for research positions today, some of them are so tightly wound up from years of intense competition that they have difficulty relaxing enough to be creative. Their lives are written in advance, unless they were born rich, because they are committed to paying back astonishing tuition debts. They'll get to learn how to live later, after tenure, or after the startup gets sold.

Long after I moved away from New Mexico, Ellery went back to school in his eighties and earned a PhD from the same university. His dissertation was about the physiology of female athletes.

Pixels in Real Life

I still remember when a professor taught me the word "pixel" (for "picture element") and how strangely awkward and fresh that esoteric term sounded, even to him. It had been in use for a decade, mostly as a way to talk about data from satellites, but almost no computers actually supported interactive pixels.

When the university took delivery of a prototype of a pixel-painting computer called a Terak, I stayed up through the night programming it with psychedelic mandalas to watch in the dark. It was hard back then to figure out an algorithm to get a mathematical function to draw fast enough to animate. I'd sneak girls into the basement of the math building to be mesmerized until dawn. A way to impress without having to talk.

I was used to encountering a medium viscerally. The oscillators and filters in the Moog synthesizer moved me in a particular way. You could feel them. Also the speakers in the lab, made of nice teak casing with a woolly face. These materials didn't change the sound, or at least not much, but a speaker was about more than the sound. A speaker was a whole object; you saw it and felt it, and you weren't forced to think of sound as something abstracted away from the rest of reality.

Everything else in the world had a presence, even high-tech things, but this interactive computer with a screen was different.

The pixels were hard and remote in the glass. The first time I powered up that Terak, I stared at them for a long time, doing nothing, trying to feel them. It wasn't that they were hard or insulated by the glass, it was that they were so abstract. There was nothing particular about a pixel. I didn't know how to be creative with faux atoms that had no intrinsic character. But I was determined to dive in.

A professor asked me to work on a National Science Foundation grant to create interactive teaching software for mathematics. This was thrilling. It paid more than the goat herd while it lasted, and I got to go to a big conference to show off my work. I programmed a little fireworks display on the Terak's screen to reward students who made it through the lessons.

Stacks

The ugly metal shelves of the NMSU library banked up against cinder-block walls. There were grooves and scratches on broad tan floor tiles, and echoes from even the tiniest sound, so you were aware of everyone else present. Good place to hide out. I spent so much time in there; I still remember the coolest, most ignored sections.

There was a corner with weird art journals from New York. Grainy photos of naked performance artists, more provocative that they would have been if the images had been clear; badly typeset poetry, also enchanting because it was not quite decipherable. The brazen crudeness of conceptual art publications from the 1970s exuded cool. It was infuriating to know that libraries in New York or San Francisco got the same stuff a half year earlier.

But mostly I was in awe. There were scores of ancient music and journals about weird geometry. The science and math sections were the strongest in the library, and oh man, they sent me over the edge. (I was a Coxeter fanatic.)*

Some of the earliest nontechnical books about computing were split in two. One half would be about a systems approach to reality and the human future. That part would be nerdy. The other half would be about the personal experience of computing. It would be ecstatic, brimming with revelations.

---

* Harold Scott MacDonald "Donald" Coxeter was the great geometer of the twentieth century. He explored the majestic domain of symmetrical forms of which geodesic domes are only a first peek. Aside from his stature in mathematics, he directly inspired not only geodesic dome architect Buckminster Fuller, but the artist M. C. Escher.

One example was *II Cybernetic Frontiers*, by Stewart Brand. The first half was an interview with Gregory Bateson* about how cybernetics would change society and the way we know the world. The other half was devoted to Spacewar!, the first networked videogame, and the fanatical devotion the game inspired.

Another example was cloaked as a grainy retort to the craft of printing, just like New York conceptual art 'zines. This was *Computer Lib/ Dream Machines*, by Ted Nelson. Unreadable in sections, thanks to an infinitesimal font, it was a glimpse of a promised land through distant fog, enchanting. It had two front covers. One cover was for a book about how computers would inspire utopian politics, the specifics of which were either not articulated or not legible. Flip it over and twirl it upright, and there was found a montage of tales and images suggesting a digital psychedelic destiny. The effect was attractive, but puzzling.† Why promote a popular revolution in culture and society in an impenetrable package?‡

These books revealed a split in the earliest days of computing culture that has never gone away. There's a big picture way of thinking about computing and a personal one.

I prefer the personal one. It's fun. The big picture approach to computing tends to foster utopian fantasies, so it's dangerous.

Burrowing in the back reaches of the library recalled my explorations of the shelf of books and records at my little school in Juárez so long ago. I wondered if I might find a new treat on a par with *The Garden of Earthly Delights*.

Prize

It was camouflaged, hidden in the most boring possible academic journal. I had finally stumbled upon Ivan Sutherland's amazing work.

These days, I am sometimes called the father of VR. My usual retort is

---

* Bateson was an anthropologist as well as one of the most prominent philosophers of cybernetics. I can't possibly summarize his work here, but I will say that he charted a way out of the terrifying vision revealed by Wiener. He suggested a humble approach to technology, in which people don't think of themselves as being placed above nature, but embedded within a larger system.

† Decades later, Ted would say that his greatest regret might have been the font size in that book.

‡ Okay, snide reader, I'll say it for you. I get complaints that my books are too hard to read. So many big words, even as I am apparently challenging digital elitism. I don't have a perfect answer to this criticism. One must write as who one is.

that it depends on whether you believe the mother of VR. VR was actually birthed by a long parade of scientists and entrepreneurs.

Ivan had started up the whole field of computer graphics with his 1963 PhD thesis, which was called Sketchpad. It was the first demo of how people could work with computer-generated images on a screen.

Sketchpad was different from the device you're probably reading this on. It had no pixels, for instance; they weren't widely used until later.

Instead of pixels, the electron beam that would normally scan back and forth to create the rows that made up the image on an old-fashioned cathode ray tube TV was hijacked. It was guided around the screen the way a hand guides a pencil, directly sketching lines that formed stick figure and outline images (just like in my haunted house displays).

With this meager foundation, Ivan not only invented, but built out one of the main avenues of human experience in our times: interaction on a screen. The impact was spectacular. It's often called the best computer demo of all time.*

A little later, in 1965, Ivan proposed a head-mounted display, which he called "the ultimate display," and then in 1969 he actually built one, known today as the Sword of Damocles, though that was actually the name of the armature that hung down from the ceiling to hold the goggles up. Through them, you could see from within a world sustained by computer programs. Ivan spoke of the "virtual world"—the art theorist Susanne Langer's term—as the place you saw through his headset.

Already I can hear the VR mavens kvetching. There is not one single detail of the story of VR that doesn't involve a priority dispute. VR still feels a little like a new gigantic uncharted territory, summoning one's inner conquistador. Everyone who becomes involved wants to coin their own new term or stake out priority to bear their memory. That means there's a lot of affront to go around and it can seep even into the telling of our primordial history.

---

* There is another contender, which is Doug Engelbart's famous first demonstration of productivity software in 1968. Doug showed text editing, windows, pointing and selecting things on the screen, collaborative editing, file versions, video conferencing, and many other designs that have become building blocks of our lives. Sometimes Ivan's demo is called the "best demo ever," while Doug's is called "the mother of all demos," even though Ivan's was earlier.

Second VR Definition: A simulated new frontier that
can evoke a grandiosity recalling the Age of Exploration or the Wild West.*

This book conveys my personal perspective; it doesn't attempt a comprehensive history or survey of ideas. Even so, I'll try to be fair.

In the case of the earliest display headsets that resemble the ones used in VR, Philco indeed built devices a few years earlier for telepresence (meaning advanced remote control robots), and the wonderful Mort Heilig[†] made devices for stereo film viewing—not to mention a number of radical artists who had put TVs in helmets in the fifties as ironic comments on how society was perhaps getting a little too absorbed in the emerging pop culture of TV.

All of this preceded Ivan's work, but none of it involved synthesizing an interactive alternate world of unlimited variations that compensated for head motion (to create the illusion that it was stationary, outside of the person), so for me, Ivan built the first headset that counts as a VR device.

Ivan's work was hidden in plain sight. He didn't have the flamboyance of a Marshall McLuhan, but he probably had more influence on the future of media than anyone else working in the 1960s. You had to read past the surface, since he presented his work with deceptively dry affect.

I love recalling the first passes of computer science because then you can see how the whole of computation is an act of invention.

---

* Hope it's okay to include a snarky definition. Snark is one of those qualities that looks better on younger people. As you get older, snark starts to come off as "old fart syndrome" even if you are no more snarky than you used to be. Am in the process of snark quotient self-assessment. For now, I can only hope the level is just right in this book.

† Mort built a few prototypes of what he called Sensorama arcade machines starting around 1962. You'd put in your quarter, step into a booth, and ease your eyes into a stereo viewer. The device played not only a stereo movie but also a stereo audio track; it shook you, blew wind at you. One of the experiences was riding a motorcycle. My favorite, though, was a "date" in which you accompanied a teenage girl on a tour of amusement park rides. There's something about that production that captured a feeling of early-1960s innocence. The guts of a Sensorama booth were jammed with projectors, tape recorders, blowers, motors. Mort had to nurse the machines to make them work. In retirement, he designed and machined his own line of scooters. Mort would go to flea markets to sell them. He told me he loved commerce. "I bring a little of myself to people, they bring me so much." I cried, remembering Mort, when my daughter grew old enough to receive one of his scooters, long after he had passed away.

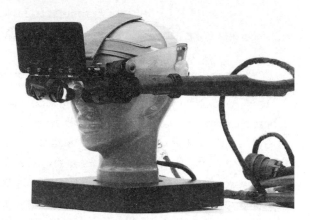

One of Ivan's earliest VR
headsets, from the late
1960s.

Ivan won the Kyoto Prize in
2012.

Nothing about computers is inevitable. But we've put such a massive number of bits into place that it's often too much work to remember how each brick of the edifice we live in is nothing but a peculiar obsession someone else put into place, once upon a time.

Pestering Strangers About Ivan

Here was the thought experiment I had imagined so long ago as a child, but now actionable: *The Garden of Earthly Delights* looking out on disciplined Texas instead of histrionic Mexico. Reading about Ivan's work was challenging for me, because each sentence took me by storm and I had to rest before moving on. Maybe the experience was made more intense because I had no way to even see a film of Ivan's demos in action, I could only imagine them.

Third VR Definition: Hope for a medium that could convey dreaming.

I was an intense dreamer. Often I would find myself taking on the identity of a cloud rolling over a mountainside or of a mountainside itself feeling villages spread on its skin over the course of centuries, the stone cathedrals pressing into my flesh while the farmers tickled me. I dreamed fantastic things that were impossible to describe. The shared world, the world out there with the other people, felt sluggish, limp, and inflexible. I longed to see what was inside the heads of other people. I wanted to show them what I explored in dreams. I imagined virtual worlds that would never grow stale because people would bring surprises to each other. I felt trapped without this tool. Why, *why* wasn't it around already?

At the time, actual virtual worlds were still made of outlines and stick figures, just like in Sketchpad. Simple things: a grid to indicate the floor, simple geometric forms.

But just recently, filled-in still images of three-dimensional objects had finally been rendered by computers. Not moving or interactive images, but even so: There, in a computer science journal, was an image of a cube with shaded sides! Computers were like toddlers learning to crayon between the lines.

Just a cube, but it was made by a computer, and computers would become more and more powerful. Someday trees and skies, creatures and seas. So someday not just *The Garden of Earthly Delights*, but Any Garden!

You would eventually be able to make any place and be in it via this device that hung from the ceiling. Plus, other people could be in there with you, just like in the networked video game Stewart Brand wrote about, but not restricted to spaceships. It was such a powerful thought that blood rushed to my head and I had to sit on the floor for a while.

I was immediately obsessed with the potential for multiple people to share such a place, and to achieve a new type of consensus reality, and it seemed to me that a "social version" of the virtual world would have to be called virtual reality. This in turn required that people would have bodies in VR so that they could see each other, and so on, but all that would have to wait for computers to get better.

I was fifteen years old and vibrating with excitement. I had to tell someone, anyone. I would find myself running out the library door so that I didn't have to keep quiet; rushing up to strangers on the sidewalk out in the hard New Mexico sunshine.

"You have to look at this! We'll be able to put each other in dreams using computers! Anything you can imagine! It's not just going to be in our heads anymore!" I'd then wave a picture of a cube in front of a random, poor soul, and that person would politely navigate around me. Why were people so blind to the most amazing thing happening in the world?

(Please remember that this was before the Internet, so to talk to a stranger you had no choice but to walk up to one.)

4. Why I Love VR (About the Basics)

Recall that this book has two kinds of chapters. One kind tells a story and the other explores topics in virtual reality. This is the first of the second kind of chapter. Here I will introduce some general ideas about VR. The other "about" chapters to come will examine different aspects of a VR system, such as visual displays.

The Mirror Reveals

Even though it's finally becoming more widely accessible, a lot of the joy in VR remains in just thinking about it.

One way to think about VR is through surreal thought experiments. Imagine the universe with a person-shaped cavity excised from it. What can we say about the inward-facing surface that surrounds the cavity?

Fourth VR Definition: The substitution of the interface between a person and the physical environment with an interface to a simulated environment.

You can think of an ideal virtual reality setup as a sensorimotor mirror; an inversion of the human body, if you like.

In order for the visual aspect of VR to work, for example, you have to calculate what your eyes should see in the virtual world as you look around. Your eyes wander and the VR computer must constantly, and as instantly as possible, calculate whatever graphic images they would see were the virtual world real. When you turn to look to the right, the virtual world must swivel to the left in compensation, to create the illusion that it is stationary, outside of you and independent.

Back in the early days I used to luxuriate when I described this most basic principle of VR to people who had never heard of it. People just flipped out when they first got it!

Wherever the human body has a sensor, like an eye or an ear, a VR system must present a stimulus to that body part to create an illusory world. The eye needs visual display, for instance, and the ear needs an audio speaker. But unlike prior media devices, every component of VR must function in tight reflection of the motion of the human body.

Fifth VR Definition: A mirror image of a person's
sensory and motor organs, or if you like,
an inversion of a person.

Or, to make it more concrete:

Sixth VR Definition: An ever growing set of gadgets
that work together and match up with human
sensory or motor organs. Goggles, gloves, floors
that scroll, so you can feel like you're walking far in
the virtual world even though you remain in the same
physical spot; the list will never end.

The ultimate VR system would include enough displays, actuators, sensors, and other devices to allow a person to experience, well, anything. Become *any* animal or alien, in *any* environment, doing *anything*, with effectively perfect realism.

Words like "any" show up a lot in VR definitions, but after working with VR, most researchers learn to be suspicious whenever "any" is uttered. What's wrong with this innocent-seeming little word?

My position is that in a given year, no matter how far we project into the future, the best possible VR system will never achieve complete coverage of all the human senses or measurement of everything there is to be measured from a person. Whatever VR is, it's always chasing toward an ultimate destination that probably can't ever be reached. Not everyone agrees with me about that.

Some VR freaks think that VR will eventually become "better" than the human nervous system, so that it wouldn't make sense to try to improve it anymore. It would then be as good as people could ever appreciate.

I don't see things that way. One reason is that the human nervous system benefits from hundreds of millions of years of evolution and can tune itself to the quantum limit of reality in special cases already. The retina can respond to a single photon, for instance. When we think technology can surpass our bodies in a comprehensive way, we are forgetting what we know about our bodies and physical reality. The universe doesn't have infinitely fine grains, and the body is already tuned in as finely as anything can ever be, when it needs to be.

There will always be circumstances in which an illusion rendered by a layer of media technology, no matter how refined, will be revealed to be a little clumsy in comparison to unmediated reality. The forgery will be a little coarser and slower; a trace less graceful.*

But that's not even the best reason to think that our simulations will not surpass our bodies.

When confronted with high-quality VR, we become more discriminating. VR trains us to perceive better, until that latest fancy VR setup

---

* Arguments about this point of view were common—voracious, actually—in the 1980s. The counter was and remains that we will eventually gain mastery over physical reality in every detail, through a hypothetical, ultimate nanotechnology, so that there will no longer be a distinction between virtual and physical reality. For instance, what about a hypothetical future of augmented human anatomy? If we come to see the world better through enhanced sensory organs, would we not also be able to feed those same organs directly with data from a simulation? These arguments go round and round, but I still think the brain will just get better and better at detecting forgeries. Remember, we can't outrun the *interactivity* of reality. If we someday enhance our vision with super-high-resolution artificial retinas that can see many more colors—even then, the key to perception will be the interactivity, the probing. Even then, VR won't look as real as what we can see of physicality through our new eyes, provided we allow those eyes to roam true.

doesn't seem so high-quality anymore. The whole point of advancing VR is to make VR always obsolete.

Through VR, we learn to sense what makes physical reality real. We learn to perform new probing experiments with our bodies and our thoughts, moment to moment, mostly unconsciously. Encountering top-quality VR refines our ability to discern and enjoy physicality. This is a theme I will return to many times.

Our brains are not stuck in place; they're remarkably plastic and adaptive. We are not fixed targets, but creative processes. If time machines are ever invented, then it would become possible to snatch someone from the present and put that person in a future, highly sophisticated VR setup. And that person would be fooled. Similarly, if we could grab people from the past and put them in our present-day VR systems, *they* would be fooled.

To paraphrase Abraham Lincoln: You can fool some of the people with the VR of their own time, and all of the people with VR from future times, but you can't fool all of the people with the VR of their own time.

The reason is that human cognition is in motion and will generally outrace progress in VR.

Seventh VR Definition: A coarser, simulated reality fosters appreciation of the depth of physical reality in comparison. As VR progresses in the future, human perception will be nurtured by it and will learn to find ever more depth in physical reality.

Because of future progress in VR technology, we humans will become ever better natural detectives, learning new tricks to distinguish illusion from reality.

Both today's natural retinas and tomorrow's artificial ones will harbor flaws and illusions, for that will always be true for all transducers. The brain will constantly twiddle and test, and learn to see around those illusions. The unceasing flow of tiny learning forces—pressed finger against pliant material, sensor cell in the skin exciting a neuron that signals the brain as the pressure reflects—this flow is the blood of perception.

Verb Not Noun

Virtual reality researchers prefer verbs to nouns when it comes to describing how people interact with reality. The boundary between a person and the rest of the universe is more like a game of strategy than like a movie.

The body and the brain are constantly probing and testing reality. Reality is what pushes back. From the brain's point of view, reality is the expectation of what the next moment will be like, but that expectation must constantly be adjusted.

A sense of cognitive momentum, of moment-to-moment anticipation, becomes palpable in VR.*

So how can we simulate an alternate reality for a person? VR is not about simulating reality, really, but about stimulating neural expectations.

Eighth VR Definition: Technology that rallies the brain to fill in the blanks and cover over the mistakes of a simulator, in order to make a simulated reality seem better than it ought to.

Actionable definitions of VR are always about the process of approaching an ideal rather than achieving it. Approach, rather than arrival, is what makes science realistic, after all. (If that way of understanding science isn't clear to you, please read this footnote.)†

---

* It's been suggested that it's the same thing as the "chi" in "tai chi," but I don't know enough to comment on that.

† Here is an example showing how science is about approach instead of arrival: The twentieth century brought us two theories of physics, quantum field theory and general relativity, that are so good that no one has yet devised an experiment that exposes an inaccuracy in either of them. And yet they disagree with each other when it comes to some extreme situations related to the universe as a whole, or to black holes.

So we know physics isn't "done." That doesn't mean that progress hasn't been authentic: Relativity gives our GPS sensors accuracy and quantum field theory allows us to stuff the resulting data into fiber optic cables under the sea. We couldn't do any of that without the theories. And yet there's obviously more to be discovered.

Science isn't about the certainty of coming to a final conclusion, and this can make it emotionally unsatisfying. The mind thinks thoughts, so it wants reality to be like a thought, to stake out a position, to be Platonic. But science is only about making gradual progress, holding a candle in a great darkness.

Minds can get stubborn about thoughts, and expect reality to be a certain way and have that be the end of it. Alas, eternal reality hasn't been revealed to us totally and instantly.

There's a grandeur in the gradual way science progresses. It takes a while to get used to it, but once you see it, the incremental ascent of science becomes a thing of beauty and a foundation for trust.

I appreciate the infinite elusiveness of a perfected, completed form of VR in the light of this sensibility. Reality can never be fully known, and neither can virtual reality.

Ninth VR Definition: The investigation of the sensorimotor loop that connects people with their world and the ways it can be tweaked through engineering. The investigation has no end, since people change under investigation.

A Vice to Avoid

An obstacle to understanding is that popular metaphors for the nervous system come from commonplace gadgets that operate on principles that are alien to the brain. It is quite common, for instance, to think of eyes as being like cameras, ears like microphones, and brains like computers. We imagine ourselves as USB Mr. Potato Heads.

A better metaphor: The head is a spy submarine, sent out into the world to perform a multitude of experimental missions to try to discern what's

---

Since science isn't absolutely done, people can feel emotionally cheated by science. It's like when we want a perfect king, and yet all we can ever really get is an imperfect politician. It sucks.

I feel these emotions too. Sometimes I wish science could be perfect. But you just have to get used to the way things work in this reality of ours. It's a miracle, an amazement, a stunning blessing, that we can make progress at all. We can understand more than we used to. Even so, it sucks that we are still not omniscient.

The imperfection of our understanding can make us lash out at science the way we lash out at our politicians. Climate change deniers and antivaccine people argue that if science isn't done, then *nothing* has been settled. Sometimes, certain AI people can believe that just because we've learned a few things about how brains work, we must already understand *everything* crucial about how brains work.

I feel the emotions behind these exaggerations, but what makes science worth trusting is that it doesn't promise everything. Only charlatans promise everything. Science has settled *some* issues. Boy, is it hard to accept *some* when what you really want is *everything*.

When you lash out at decent but imperfect politicians, you only get worse politicians who pretend to be kings. When you lash out at incomplete but valid science, you offer yourself up to con artists.

out there. A camera placed on a tripod typically takes a more accurate picture than one held by hand. The opposite is true for eyes.

If you immobilize your head in a vise, and to complete the picture, if you inactivate the muscles that move the eyes about in their sockets, you will have simulated putting your eyes on a tripod. For a moment you'll continue to see as before, though it might feel as if you're looking at a movie. Then something terrifying will happen. The world around you will fade to a sickly gray and then disappear.

Vision depends on continuous experimentation carried out by the nervous system, actualized in large part through the motion of the head and eyes. Look around you and notice what happens as you move your head in the smallest increments you can manage. Seriously, stop reading for a moment and just look around and notice how you see.*

Move your head absolutely as little as you can, and you will still see that edges of objects at different distances line up differently with each other in response to the motion. This is called "motion parallax" in the trade. It's a huge part of 3-D perception.

You will also see subtle changes in the lighting and texture of many things. Look at another person's skin and you will see that you are probing into the interior of the skin as your head moves. (The skin and eyes evolved together to make this work.) If you are looking at another person, you will see, if you pay close attention, an unfathomable variety of tiny head motion messages bouncing back and forth between you. There is a secret visual motion language between all people.

If you are not able to perceive these things, try going into VR for a while and then come out and try again.

Vision works by pursuing and noticing changes instead of constancies, and therefore a neural expectation exists of what is about to be seen. Your nervous system acts a little like a scientific community; it is voraciously curious, constantly testing out ideas about what's out there. A virtual reality system succeeds when it temporarily convinces the "community" to rally behind an alternate hypothesis. (If VR ever succeeds on a permanent basis, we will have entered into a new form of catastrophic political failure. The more we each become familiar with successful *temporary* VR experiences, however, the less vulnerable we become to this bleak fate.)

---

* If you're blind, the principle works equally for hearing.

Once the nervous system has been given enough cues to treat the virtual world as the world on which to base expectations, VR can start to feel real, realer than it ought to, in a way, which is a dead giveaway.

The nervous system is holistic, so it chooses one external world at a time to believe in. A virtual reality system's task is to sway the nervous system over a threshold so that the brain believes in the virtual world instead of the physical one for a while.

Tenth VR Definition: Reality, from a cognitive point of view, is the brain's expectation of the next moment. In virtual reality, the brain has been persuaded to expect virtual stuff instead of real stuff for a while.

The Technology of Noticing Oneself

VR is a hard topic to explain because it's hard to contain. It directly connects to every other discipline. I've had visiting appointments in departments of math, medicine, physics, journalism, art, cognitive science, government, business, cinema, and sure, computer science, all because of my work in this one discipline of VR.

Eleventh VR Definition: VR is the most centrally situated discipline.

For me, VR's greatest value is as a palate cleanser.

Everyone becomes used to the most basic experiences of life and our world, and we take them for granted. Once your nervous system adapts to a virtual world, however, and then you come back, you have a chance to experience being born again in microcosm. The most ordinary surface, cheap wood or plain dirt, is bejeweled in infinite detail for a short while. To look into another's eyes is almost too intense.

Virtual reality was and remains a revelation. And it's not just the world external to you that is revealed anew. There's a moment that comes when you notice that even when everything changes, you are still there, at the center, experiencing whatever is present.

After my hand got giant, it was natural to experiment with changing into animals, a splendid variety of creatures, or even into animate clouds. After you transform your body enough, you start to feel a most remarkable effect. Everything about you and your world can change, and yet you are still there.

This experience is so simple that it is hard to convey. In everyday life we become used to the miracle of being alive. It feels ordinary. We can start to feel as though the whole world, including us, is nothing but mechanism.

Mechanisms are modular. If the parts of a car are replaced one by one with the parts of a helicopter, then afterward you will end up with either a helicopter or an inert meld of junk, but not with a car.

In virtual reality you can similarly take away all the elements of experience piece by piece. You take away the room and replace it with Seattle. Then take away your body and replace it with a giant body. All the pieces are gone and yet there you are, still experiencing what is left. Therefore, you are different from a car or a helicopter.

Your center of experience persists even after the body changes and the rest of the world changes. Virtual reality peels away phenomena and reveals that consciousness remains and is real. Virtual reality is the technology that exposes you to yourself.

There's no guarantee that a tourist in VR will notice the most important sight. I did not notice this most basic aspect of what I was working on until I experienced bugs in VR, like the giant hand. I wish I knew what threshold of elements might bring other people to appreciate the simplest and most profound quality of the VR experience.

Twelfth VR Definition: VR is the technology of noticing experience itself.

As technology changes everything, we here have a chance to discover that by pushing tech as far as possible we can rediscover something in ourselves that transcends technology.

VR is the most humanistic approach to information. It suggests an inner-centered conception of life, and of computing, that is almost the opposite of what has become familiar to most people,* and that inversion has vast implications.

VR researchers have to acknowledge the reality of inner life, for without it virtual reality would be an absurd idea. A person's Facebook page can continue after death, but not the person's VR experience. Who is the VR experience for, if not for you?

VR lets you feel your consciousness in its pure form. There you are, the fixed point in a system where everything else can change.

From inside VR you can experience flying with friends, all of you transformed into glittering angels soaring above an alien planet encrusted with animate gold spires. Consider who is there, exactly, while you float above those golden spires.

Most technology reinforces the feeling that reality is just a sea of gadgets; your brain and your phone and the cloud computing service all merging into one superbrain. You talk to Siri or Cortana as if they were people.

VR is the technology that instead highlights the existence of your subjective experience. It proves you are real.

---

* This is one of those ideas that seems so obvious to some people that they find it tedious if I elaborate, while others find it puzzling. If it seems puzzling, you might sneak a peek at the sections about artificial intelligence later in the book, starting with "Birth of a Religion" on page 257.

5. Bug in the System (About the Dark Side of VR)

Paranoid Android

After my mother died, sequences of words had shown me the way out of the hospital. "Choose life." Now that I was a teenager, other sequences of words nearly sent me back in.

Ellery had given me his copy of Norbert Wiener's *The Human Use of Human Beings* when I started college and became interested in computer science. This is a profoundly terrifying book, written so early in the game that Wiener had to define basic terms. He articulated an approach to the future of computation that he called cybernetics.

Wiener realized that someday, when computers would become thoroughly integrated into human affairs, we would only be able to understand people and computers as portions of a system that included both. This might seem obvious, but it was a giant leap of prescience at the time.*

Wiener was not a popular guy at the dawn of computation. He was outlived by various critics, who described him to me in unkind terms. Whatever he was like as a person, and I have no opinion about that, he had the clarity of mind that comes most easily to whoever first arrives at a new terrain of ideas.

---

* Wiener's work was the beginning of "systems" writing about computation but was too early to have an accompanying psychedelic half, like *II Cybernetic Frontiers* or *Computer Lib/Dream Machines*.

One reason the term "artificial intelligence" came into being (at a conference at Dartmouth in the late 1950s) is that more than a few of Wiener's colleagues couldn't stand him. They felt compelled to come up with an alternative name, because cybernetics was starting to catch on and was associated with Wiener. The alternative they came up with did not mean the same thing.

"Artificial intelligence" purported to describe qualities of future computers without reference to people, suggesting that computers will become free-standing entities that will exist even if all the people died; even after no one is left to observe them.

"Cybernetics," by contrast, proposed only that computers and people would have to be understood in the context of each other. It wasn't concerned with metaphysics.

Wiener was right: AI muddies the waters. I'll get back to thoughts about AI toward the end of the book, but in the meantime, let's consider what Wiener's thinking might mean for VR.

Terror Equation

An equation to summarize what makes Wiener's book terrifying:

$$\text{Turing}^{\text{Moore's Law}} * (\text{Pavlov, Watson, Skinner}) = \text{Zombie Apocalypse}$$

World War II left a sense of dread in its wake; a feeling that human agency might come under the threat of technology. The Nazis had harnessed new technologies like cinematic propaganda to make a large population complicit in the invention of an industrialized version of genocide. One tiny pixel in the gray was my mother, a most unlikely survivor.

After the war, everyone wondered how this could have happened. Could it happen again? Would we recognize the early stages? What to do then?

The postwar period was haunted by the fear of mind control. Psychologists like Ivan Pavlov, John B. Watson, and B. F. Skinner had shown that the application of controlled feedback could be used to modify behavior. That dark metallic flavor of modern paranoia that is so familiar in the work of William Burroughs, Thomas Pynchon, Philip K. Dick, the cyberpunk school—indeed in most modern science fiction; it all started with nonscientists freaking out when a few scientists bragged about power trips in the lab.

Some of the original behaviorists dripped with arrogance—as if they had the right to declare how other people could be engineered in the lab or in society—as well as totalism; a sense that no other approach to studying people would ever matter.

Pavlov was the guy who rang a bell when a dog ate and then proved he could get the dog to salivate by the bell alone. Watson was the guy who did the cruel "Little Albert" experiment. He scared a baby when animals were around in order to prove he could make a human become scared of animals forever. And Skinner formalized an experimental box for conditioning animals in laboratories.

Behaviorism has been reduced to gadgetry in pop culture. You tweet for an instant treat, for attention, even if you're the president. You salivate when you hear a dog whistle. The Skinner box was the archetype. A person in a Skinner box has an *illusion of control* but is actually controlled by the box, or really by whoever is behind the box.

A crucial distinction has to be made, and it isn't easy to get it right. I was repulsed by the *culture* of behaviorists, though not behaviorism itself, which can be useful science. Back in the day I thought about goat training as an example of the useful side of behaviorism, but today I might bring up cognitive behavioral therapy.

In college, I became obsessed with the problem of how to draw the line between useful science and creepy power trips. Thoughts raced through my head, keeping me up through the night. "We need science in order to survive; plus it's beautiful. But scientists can be creeps. Science as dramatized and motivated by creeps can do terrible damage. How can we do science when we're not always good enough to deserve it?"

What upset me most about behaviorism was the glaze of antihuman feeling; publicity seeking through sadistic experiments. You can use any technology to dramatize new forms of cruelty. But why?

Behaviorism was not unique as a paranoia factory. Genetics is useful and valid, and yet geneticists had from time to time veered into eugenic utopian thinking that was horribly antihuman and evil. Scientists contributed to the murder of my relatives and the imprisonment and torture of my mother, along with millions of others.

If you want to get a feeling for the paranoia still in play when I first studied computer science, I recommend the original version of the movie *The Manchurian Candidate*. An American soldier is brainwashed, not through propaganda, Stockholm syndrome, or any other ploy within the scope of human interactions. Instead, the soldier is subjected to mercilessly algorithmic, sterile stimulus and feedback. Skinner-like mind hacking was portrayed again in *A Clockwork Orange* and in too many other novels and films to list.

What could be more awful than the possibility that some nerd is running you, without your knowledge, as if you were a character in a videogame?

Films and novels from after the war until around the turn of the century usually proposed that in order to control humans, hypnosis would play a role, or maybe a supposed truth serum. It wasn't just the movies! The CIA actually gave people LSD—without notice or consent—to see if that would facilitate mind control.

Wiener extrapolated that computers might become powerful enough to run fancier Skinner boxes, more effective, harder to detect, and infinitely creepier. Read Wiener carefully and it's clear that with good enough sensors, good enough computation, and good enough sensory feedback, a Skinner box could be implemented around a person in a waking state without that person's realizing it. Wiener comforts the reader by pointing out that it would be so hard to create a giant computational facility and communications network that the danger is only theoretical.

Bipolar Bits

A suffocating thought hit me only a few months after my initial computer graphics reveries. It was so awful that I had to unthink it right way; it burned. But then, over the years, I reencountered this black notion at odd moments, and gradually found an accord—but more about that later.

The thought is that virtual world technology is inherently the ideal apparatus for the ultimate Skinner box. A virtual world could be, precisely, the creepiest technology ever.

Remember, at this time virtual worlds were made only of Spartan monotone line renderings and only viewed on rare occasions through huge, industrial-scale rigs in a few labs.

But my daydreams, and probably my dreams at night, were filled with imagining what this new technology would be like. It would be beautiful, expressive, sensitive. It would be Hieronymus Bosch crossed with Bach crossed with chocolate. My hand would be measured, would turn into an unconstrained appendage, maybe a hand, maybe a wing. I'd fly through the Mandelbrot Set on a date, I'd program through dancing; make music with my friends by growing imaginary plants.

The terror arose from one word in that paragraph, and it is "measure."

Wiener considered how computers could fit into the world. Up to that time, computers had been mostly used in rather abstract, formal ways, to break secret codes or calculate missile trajectories. Stacks of punch cards handed to a technician behind a window. In cases like that, there is a discrete moment when an operator enters input data into the computer, such as an encrypted enemy message, and then runs the program, and then reads the output. Indeed, the formal definitions of computation from Turing and von Neumann were first expressed around this model of discrete input, processing, and output stages.

But what if computers were running all the time, interacting with the world, embedded in the world? This is exactly what Ivan Sutherland had prototyped!

"Cyber" comes from the Greek and is related to navigation. When you sail, you must constantly adjust to changes in wind and surf. In the same way, a computer would have sensors to measure the world and actuators to affect the world. A computer embedded in the world would be a little like a robotic sailor even if it was fixed in place. Maybe it could only look out of cameras, receive text keyboards, and then put images up on a screen or perhaps control machinery. Thus, "cybernetics."

This vision of computation was portrayed in *2001: A Space Odyssey*. HAL doesn't reside in an android that walks around; it just sits there, ambient. Nonetheless it sails. It navigates both the spacecraft and what goes on inside.

Now consider a Skinner box. What are the components? There is measurement of the creature in the box. Did the rat press the button? There is feedback. Will food appear? What causes action to be triggered by measurement? In the original experiments, a live scientist was at the controls, but these days, it's an algorithm.

The components of a Skinner box and those of a cybernetic computer are essentially the same. This is perhaps too elementary an observation even to state at this late date, but when I was young the connection was still fresh and shattering.

Virtual reality, in order to work well, would have to include the best possible sensing of human activity ever. And it could create practically any experience as a form of feedback. It could turn out to be the evilest invention of all time.

Thirteenth VR Definition: The perfect tool for the perfect, perfectly evil Skinner box.

Wait, don't think that! Pull back! Think of anything else. Learn the shakuhachi, travel to exotic places, avoid the Thought.

6. Road

Dome Done

Seventeen years old was I, and the dome was at long last done enough to be called done. I had almost completed a bachelor's degree in math and was already a teaching assistant in graduate classes.

But I was afraid that I might be falling into a trap, learning to build evil machines. I had to see more of the world, get perspective.

As if on cue, I met a fellow who was a few years older and described himself as a poet from New York. I had never met someone who self-reported in that fashion. He had long hair and a goatee. He went to an art school in the countryside outside the big city.

I suddenly had to go there. Why? In part it was the allure of the avant-garde 'zines that I read in the library. In part it was my fascination with Conlon Nancarrow,* synthesizers, and experimental music. No, none of that was it. My parents had been artists in New York. They had the bug at one time. I had to go there, to retrace my mother's steps.

---

* Conlon Nancarrow was a composer in Mexico City. I describe my relationship with him in *Who Owns the Future?* He had been born American, but was refused re-entry on the grounds that he was "prematurely anti-fascist" after he fought with the Abraham Lincoln brigade in Spain in World War II. Conlon hand-punched player piano rolls in order to achieve complete freedom and accuracy in the time domain, so he was a pioneer of exploring the meaning of limitlessness in art. If you want to listen, see if you can find the old 1750 Arch LP recordings. The later digital recordings are a little dry and seem to me to miss the point.

The money was a big problem. This school was walloping expensive compared to NMSU. My father took out a loan, with the dome as collateral.

We drove across the country—the guy had a van. I was astounded by how steamy green the terrain became as we raced eastward. When Manhattan came into view, I became so excited that it was like a seizure. We didn't stop there, but went on to the little upstate campus.

I was absolutely unprepared for the snobbery. Almost all the kids were from wealthy families. I had read Thorstein Veblen, a favorite of my father's, and he wrote the script these kids lived by. Every expression was a complaint. "Born too late," went a student-penned folk song. Poor us, we missed the sixties.

There was spectacular, showy waste. Esoteric shiny sports cars wantonly junked up in purposeful accidents on a Friday night. A story for Saturday.

But there was also a pervasive affectation to suffering and poverty. The dorms were bombed out, simulacra of the way much of New York City—the poor areas—looked in that era. It was trendy to live low, to live punk. Everyone was a radical; everyone knew more than the others about real life, real poverty, real suffering.

The richest kids got hooked on heroin. It was accepted. They obligingly enacted personality cults for one another. One was a great poet, this other one a great filmmaker.

I don't think there was another kid at the school who had ever had to earn a living. But I so wanted to be accepted by them. To be treated as a real artist. Not a chance of that, of course. A red *H* for hillbilly was sewn on my skin.

Previously I was aware that I was slightly and weirdly privileged, and I was. After all, it wasn't me who drowned at the bottom of my neighborhood pool. My skin color had elevated my status a tiny but crucial bit.

But I realized that status is fractal; the pattern repeats itself at every scale, small and large. When the titans of industry are gathered in a room, there will always be one who is the designated loser, relatively speaking. When poor tough kids cluster, there's always one who's the top dog. I experienced yet other local minima.

This is not entirely fair. I met a few reasonable, levelheaded students. But overall I am telling it as it was.

Film Flam

One good thing about the place was that it gave me my first chance to learn how to talk about ideas. The students loved bull sessions in which they might sound like important intellectuals. The most common topic was cinema.

The campus was an oasis for avant-garde filmmakers: itinerant odd-ball characters who had made only a few movies, each only a few minutes long, and who were absolutely adored by the students, including me. Figures like Stan Brakhage or Michael Snow could always earn a few bucks from a visit. Screenings were held in a rusty old temporary shed, food was served later in a dive bar with a jukebox so loud you couldn't think.

(Oh god, I still shudder when I think of that music, the same songs over and over. Most people are fixated on the music that was popular when they were young. Whether the music was bad or if it was just me, a lot of mid-to-late-seventies hits sounded awful then and still sound awful now.)

We didn't just watch, we talked. Not just film, but "film culture." An intensely savored trope that was passed around, at every single avant-garde film bull session, was that someday each person would be filmed from birth until death with no gaps. All would be recorded.

I riffed on the idea with a supremacist, Borgesian flair; film would overwhelm time itself. "Nothing will be forgotten, so the present and past will become less distinct. Time will become less linear and more diffuse, spread out like a map instead of a string."

This little rant was one of the few wedges that got me on the inside of the social world, if only for a moment. The commanding novelty of the idea of total filming of everything was so alluring that it seemed like the future, the triumph of film over everything else. Cinema supremacy was the future! Basically I was flattering the hell out of everyone in earshot in a bid to be accepted.

The obscurity of weird film culture was part of the appeal. We liked knowing that ordinary lugs didn't know who Maya Deren was.

(You probably don't either, but this tiny circle of filmmakers invented the patterns and styles of a lot of what you know how to see in music videos. They ended up having influence comparable to Steven Spielberg or George Lucas. I never would have guessed.)

One day I was walking along in the preposterous humidity and a terrible thought struck. The forbidden thought. *What if I was proposing the human use of human beings?* Whiplash.

I remember asking, at one of the pompous discussions that followed a screening of a new film from the Mekas brothers or Jack Smith: "Who will be responsible for this total film of a person's life? Who will position cameras, adjust color, cut between cameras?"

"Film requires a swarm of decisions," I went on. "It's real work. If each person must direct the comprehensive film of one's own life, then there would be no time for any life. Film would choke off everything else and would create a stasis, a still image. If, instead, someone else was to direct, then fascism would result, because that someone else would control memory, and so would control everything. Therefore, we must not film everything. We must forget enough to be free."

Oddly, this argument had no takers. It was sort of paranoid and sort of neo-Marxist—the brew that young, stuck-up kids prefer; I thought it would win sympathy. Furthermore, the idea felt important and potentially true. But instead there were only sullen smirks. The thrilling comfort of the earlier flattery was invincible.

While I failed to make my point, the argument ultimately comforted me more than it terrified. I felt deeply guilty about not remembering my mother better, but I had achieved clarity that strategic forgetting can sometimes be the only path to freedom.

I didn't get so much from the classes I was officially enrolled in. The math and science was mediocre—an afterthought. No computers present, no interest in them and certainly no understanding of them. I had to put my computer science life on hold. Worse, the music faculty was occult and mean, for terrible reasons I soon came to understand.

But, before that, a nice thing happened.

First Time in the City

I would take the train down into Manhattan on weekends and stay with a friend of my parents, Ruth Morley, a costume designer for the movies. She's remembered for her work in *Annie Hall* and *Tootsie*. She lived in a purple penthouse just behind the Dakota and had two daughters who were just a little older than me.

A trace of my parents' New York life persisted and welcomed me! From that purple penthouse I roamed out and met up with the actual avant-garde music scene, not the pretend one upstate, and that was wonderful. I spent time with John Cage and other musical figures of that era. The exquisite Laurie Spiegel, a goddess of synthesizers, and the equally lovely and even more intimidating rising star Laurie Anderson.

New York City amplified you right back at yourself, a giant parabolic mirror. As you walked down the street you made eye contact and exchanged subconscious signals with thousands of people. You dove into the densest hub of fates. If you were looking for trouble, there it was. Or love, or mutual adoration, or pitiful falls from grace.

No longer so today. Everyone is looking at their phones.

There was a wealthy, elegant elderly widow, European royalty (but wasn't everyone?), who served as the stealth benefactor of much of the experimental music scene in New York. There's always someone like that behind the scenes of an apparently organic movement.

She was said to live in a house built in the form of a giant erect spike, the largest single stainless steel piece ever made, or so went the story. Her deceased husband's bones were supposedly suspended in a mobile within the spike's pinnacle.

She would gather Cage and other luminaries for crazed all-night adventures. We'd go dancing, then buzz by a rich and famous person's home, then roam back alleys to steal discarded pieces of the best cheeses from the garbage bins of the top restaurants. I would become too exhausted to keep up by around four in the morning.

New York was crime ridden in those days. The movie *Taxi Driver*, for which Ruth designed the costumes, captured it accurately. Almost everyone had been mugged. But there was a conceit in the avant-garde art world that it was all about attitude. If you approached the city with the right mind and heart, no ill would befall you. (Much later, in the 1990s, John Cage was finally mugged. He was shaken and we all were.)

A composer named Charlie Morrow organized a nutty guerrilla band that managed to invade the Stock Exchange and enliven the floor before being chased off by guards with whistles, whom we treated as band members. I made a bassoon out of sausage for one concert, and it sounded fine.

I was an intense piano player. I would bleed on the keys. Part of it was that I was trying to play Nancarrow's blistering later piano rolls by hand, which is impossible. But also I always had an almost crisis level of emotion, whatever that emotion was. When I played, it was for my life, every time.

I wish you could hear what I remember. I remember living in my own piano world, made of intense chord clashes and rhythms, giving way to almost vanishing delicate patterns. I did a lot of strange pedaling tricks, like opening and closing the dampers repeatedly for tremolo effects, or hovering them just above the vibrations well after the note had seemed to be over. My notes had afterglows. I loved crazy fast arpeggios, following Nancarrow, and had a trick where my hands would flip over to accomplish them. Whether what I remember is what other people heard at the time is hard to say.

I played piano in the Ear Inn, an ancient, sagging riverside pub, recently reborn as a composers' hangout. Another supercharged piano player, Charlemagne Palestine, would compete with me for the bench, occasionally pushing me off.

There's not much in the way of documentation from that time, no recording of how I used to play the piano. But I did have a score on the cover of *Ear* magazine, a notorious avant-garde music publication with that amazing art 'zine vibe that had spoken to me back in the library in New Mexico. Someone with the magazine would have to go up to the Dakota to beg cash from John and Yoko for each print run. My cover was a morph between a clarinet key system and the subway map of the time. A clarinet choir rode the rails to play it.

Down Spiral

Back at the upstate school, things got worse and worse. I had a job as a piano tutor. A student broke into tears one day, saying that one of the professors would grope her. A different student later sobbed that she was being forced into sex with another teacher. Then a third one had a similar story.

A male student committed suicide. He was from a particularly rich family and had lived in one of the bombed-out dorms. An untreated schizophrenic; bizarre that he was knocking around this place that couldn't care for him.

I was in a stall in the bathroom in the music department when I overheard a couple of professors laughing it up over the incident. I realized that to at least some of them, the whole place was a scam. Rich families parked idle kids who wanted to call themselves artists at the place for a lot of money. Why not partake?

What went wrong next was my fault. I was always looking for parental figures in the wrong places. It's a bad habit that arose from my mother's death that I didn't really overcome until decades later when I became a parent myself.

I needed a mentor; a parental figure. All the faculty and staff types I had approached were beyond uninterested. A lot of them had "real" positions at places like Princeton and treated the campus as an ATM, to be fled as quickly as possible after cash was withdrawn.

Maybe I should have gone to Ruth in NYC, but it seemed like such an imposition on someone who was already so generous to me. How could I tell my parents' friend I was petrified? Instead I took up with yet another campus schizophrenic (there was an oversupply), an older, failed mathematician who squatted in one building after another, harassing everyone.

I got drawn into the guy's vortex and ended up supporting his cause instead of my own. He wanted to be recognized by the school as a mathematician and hired for a faculty post. He wanted many other things. His work was gobbledygook, but I didn't see it. I lost touch with what I could have done to improve my situation. I became depressed and just flunked out. All that money, the loan! I felt that I had betrayed my father, and my mother, for that matter. I had failed on every front. My life was at an end.

The city was my happy place. Maybe I could earn enough there to pay back the loan.

But I couldn't freeload off Ruth forever. I found a reasonably well-paying gig playing clarinet in a restaurant band for a while, and shared an utterly disgusting tiny apartment in the Village with another eccentric composer/mathematician.

There's something about the 1970s that's hard to convey to younger people unless they have visited China recently. The air used to be a cauldron of poison. When you entered Manhattan, the texture of everything looked and smelled different because of the pollution.

Sometimes it was beautiful. On bad days, the buildings were rendered in charcoal and the spaces between them looked deeper and more cinematic than they do today. Sunsets could look like wounds. It felt like an alien planet. You could feel yourself being eaten away inside with each breath.

I could handle it, despite my childhood respiratory adventures. What was impossible for me was the cigarettes. Being around a cigarette makes me start to choke and black out at the same time. (Allergists tell me that the culprit is not the tobacco, but a chemical added to the paper.)

The cigarette smoke in restaurants used to get so thick that it could be hard to see across the room. As hard as I tried, I could not make the restaurant gig work.

This failure was a crossroads for me. I realized that I was metabolically unsuited to do a kind of work that I loved deeply. If I could have managed to play in places with smoke, and they *all* had smoke, then I probably wouldn't have had my careers later on in tech and science.

Manhattan was no place to be poor or depressed. The city pounced on negativity; all of it and then threw some back on you. As unthinkable as it was, I headed back to New Mexico.

Back in Flames

The New Mexico desert is primordial, striated hills worn down by time, dotted with life, illuminated in rose light; the standard of absolute, austere beauty, or if you're in a mood, just a wasted region of rocks and sand. When I returned to New Mexico, it was just dust and junk to me.

I felt dejected and lost. I couldn't very well go back to the dome. I could barely face Ellery at all, having wasted all that money from the loan. Now his salary would be drained for years to pay for my idiocy. Besides, dome life was starting to be too extreme a lifestyle even for me. But I couldn't go back to school; I had flunked out.

How to make a living? My first job was mall Santa. It was miserable. A batch of off-duty, sweaty firemen and I had to share the same thick fur Santa outfit, and it was never washed, and little kids peed on it. We choked on the smell. Our boss was one of the elves, and I was warned not to confront her about working conditions, since her brother was the district attorney. She'd come up to me and whisper-yell that my eyes weren't sparkling enough.

On the day after Christmas I went to answer an ad for road work and there were dozens of desperate, muscular men ahead of me. Even miserable manual labor was impossible to find. I finally landed a steady overnight shift at a donut shop, and felt incredibly lucky.

I eventually found a long-term rental, but it wasn't vacant yet: an ancient adobe hut in Tortugas, a village that might have become one of the pueblos, but the people never signed a treaty with the U.S. government.

With nowhere to live, and a little money saved, I decided to hitchhike to Mexico to visit Conlon again. Whenever times got rough, that's what I'd do. It was cheaper to be on the road than to stay put.

A woman who was older than me, in her twenties, told me she wanted to go along. She was married to an engineer at White Sands, though. The husband was upset.

Like all my hitchhiking adventures in Mexico, the events read as fantastical now, but they happened. I remember the bland beginning; walking with the engineer's wife to Interstate 10 to hitch to the border. My skin turned lobster red from sunburn. It hurt.

A few days later, we were chased by roving mobs of feral chihuahuas in the city of Chihuahua, then took the vertiginous train through Copper Canyon; the infinitely innocent music of the Tarahumaras played on roughhewn fiddles.

After visiting Conlon in his fancy neighborhood in Mexico City, we caught a ride through the mountains in the caravan of a roving carnival. I remember riding under the sky in the seat of a neon-green hippo-shaped vehicle meant to spin about like a carnival ride. It was barely chained, swerving and jerking, to the bed of the truck; the truck craned over sharp mountain switchbacks; and I looked straight down into terrifying tropical chasms.

A little town near Chilpancingo had fancied itself a Marxist enclave and declared independence from the federal government. A committee of young people deliberated for hours, the way Marxists do, before deciding to allow us to bunk in the commune for the evening.

Here I was in a bunk with an older woman. Was something supposed to happen? I was petrified by uncertainty.

Before I relate the next incident, I should remind the reader that I am old enough to have had to register for the tail end of the Vietnam-era draft, though no one was being called up by that time. It terrified me nonetheless. How awful it would be to be conscripted into a pointless, avoidable war; to inflict devastation on people who hadn't done anything to us?

So I had learned all about conscientious objection and the history of nonviolent action. I had taken trainings and tried on a worldview that was suspicious of the American military-industrial complex over all other things. Today I know the world is not so simple. You can't just draw a circle around what you think is the devil and declare that you've got it all figured out. That's how you eventually turn into a devil yourself, should you be successful.

Anyway, in a field near Chilpancingo the next morning, there was a confrontation between the hippielike Marxist separatists and a phalanx of uniformed federal troops, who were disciplined, standing in formation, rifles aimed.

A sudden impulse took over my brain, and I ran out in front of those rifles and yelped, in my boorish border Spanish, *¡No disparen—soy Americano!* "Don't shoot—I'm an American!"

At the time, no one would have wanted to shoot an American, but surely that fact must have sown resentment in everyone present. Aren't Americans special.

The troops didn't shoot, but I doubt they had really planned to. Later, when I turned to jelly and tremors over what I had done, I wondered where to find the line between courage and an excess of mythology.

Then, after I calmed down, we departed. Hitched a ride in a jeep with a man who claimed to be a general in the Mexican army. He had a pearl-handled revolver, a medal-encrusted chest, a high art mustache; his claim seemed plausible. His driving was terribly reckless. A blowout at racing speed, which almost sent us flying off a cliff by the ocean, terrified me more than the rifles had earlier. I think he was trying to impress the woman I was with. We offered to mount the spare tire but declined to continue on with him.

We went to a traveling circus in a little town by the sea, and at the end there were screams; a monkey had attacked and killed a small child, or so everyone said. Only much later did I realize that this was probably just a way to clear kids out of the tent so that the weary circus family could pack up and trudge to the next town. It took me almost twenty years to get that memory straight so that I was no longer plagued by the terror of it.

There was no GPS, no guidebook that could speak to most of what we encountered, no mobile phones, and no *Hitchhiker's Guide to the Galaxy* or World Wide Web. There was only you and the road, a mystery. That experience is lost now. Most travel, even "extreme" or "adventure" travel, has become a matter of selecting from a well-lit catalog. Or, worse, having algorithms select for you.

I would be less skeptical of this more orderly world of pseudoadventure if it were actually safer. Anyone attempting to retrace my steps in Mexico today might actually get shot by drug gangs, and not by choice or for any noble purpose.

The older, more mysterious world was less predictable, but also buffered, because mystery is an equalizer. If no one is sure what a stranger might be capable of, then that stranger is more likely to be left alone. The explicated world unleashes everyone to calculate risks and act.

My hut back in Tortugas was due to be available soon, so it was time to head back. In a town on the Gulf of California, on the way back, in a claustrophobic, crowded, steaming hot Mexican clothing store, my traveling companion and I suddenly changed at the same time in the little changing room, and there she was naked, along with me in front of mirrors, and yet I was so shy, so unsure, that I managed to box myself in, a little nuclear fusion chamber.

Aztec Outpost; Wheels

In a state of contained panic, we made it back home. I went to move into my adobe hut, twenty dollars a month, so excited to have my very own place for the first time, but upon entering, something was different. Hadn't there been that amazing old wood floor? "The old man who live here, he needed wood for burning. Winter was cold," said the ancient landlady with a folded, wise, weary face. What was a dirt floor to me? I moved in.

I would be awoken on cold mornings sometimes by old women selling tamales door to door, sometimes by tribal dance practice. The dance rhythm of Tortugas is oddly asymmetrical. No one not born there had ever learned to play it, or so it was said. I couldn't ever get it, and I've learned obscure music from around the world. There was an eerie traditional dancing costume that included a towering black mask embedded with mirrors, a trace of Aztec influence from precolonial days.

These days Tortugas is hard to find. It's just another patch of low-end California-like development in New Mexico. Trailer parks, convenience stores. The mountain near it does look remarkably like a turtle.

I needed a plan. Miraculously, NMSU was willing to have me around again. I got a few hours as a teaching assistant in a group theory class and did some programming for a research project. But it was not enough to live on.

Once again I looked for a job. I talked to a woman with a midwifery service for indigent farm workers. She needed an assistant but couldn't afford a real nurse, or really anyone who knew what they were doing. But I had birthed goats and got the job.

My role was not supposed to have any medical significance—I was the driver and deal-with-whatever guy—but in one case it did. A young woman was committed to a mental institution shortly after giving birth; her citizenship status was unclear. The father had just been arrested. He was attempting to smuggle something—never found out what, exactly—over the border by trying to drive his old Dodge Dart through the Rio Grande in dry season. It's kind of almost doable in places. Unfortunately, police were in pursuit, and the car was shot up. So it stayed in the mud. Meanwhile the fellow was not hit, but was in jail.

What to do with the baby? The midwife feared it would be swallowed by the system and never reunited with family. So could I take care of it? Just for a while. Sort of, you know, off the record.

Suddenly I had a baby. I would show up at the Abelian group theory seminar with this infant and bottles. You have to understand that while I had been socialized a little bit by then, I was still a cross between a juvenile hippie and a feral mountain man. The spectacle of me with a baby in a graduate math seminar must have been weird beyond measure. Fortunately, a few of the math professors were parents and helped me sort out the confusing aspects of diapers and formula.

The dad was sprung not too many days later and showed up to claim the baby. Astonishingly, he looked rather like me, also a hippie mountain man. Maybe there were a lot of us.

He turned out to be a sensitive and attentive father, and the family was reunited later and did well. But an event of great consequence for me unfolded at that moment. The dad said, "Oh man, thanks so much for taking care of my little angel. What can I do for you? Do you need a car?"

Wow, did I need a car. To have one's own car in that part of the world was to enter civilization as an equal. It meant being able to work anywhere, see anyone. What a remarkable gift! What luck!

"All you have to do is tow it out of the river. Not sure if it's on the U.S.

or Mexico side now, but I doubt anyone will give you any trouble. Go see if it's still there."

I got a guy from the feed store to tow it, and there it was, all mine. Dodge Darts with slant six engines were indestructible. Sure, the bottom had rusted out so you saw the road go by underfoot. It rarely rained, so who cared about puddles? Yes, it was important not to burn your feet on the exhaust. And, well, it started with a screwdriver, and there were bullet holes in the side.

When the dad came over to present me with the title, he helpfully pasted bumper stickers over the bullet holes. Looked reasonable.

There was no backseat, but I saw that as an opportunity. I stuffed some bales of hay back there and turned the thing into a goat limo. I picked up work moving the lovely creatures around in style.

For the first few years, the brakes would fail from time to time, so I'd have to rub the side of the car along the raised dirt that inevitably lined the horribly corrugated desert roads on which I roamed in order to bring it to a stop. One time I had to engage the side of a low stone wall surrounding a cute park in Juárez in order to stop for a red light. I certainly had no worries about damaging the car's appearance.*

That car would eventually get me all the way to Silicon Valley and my new life (though by then it had reliable brakes). I cried when I finally gave it up, after the California Highway Patrol stopped me and said, "You've got to be kidding." I think that was also when I was wrestled to the ground after being observed starting a car with a screwdriver.

Anyway, once I had a car, new vistas opened up. I had internalized the narrative of the nonviolent left and was searching for a path to significance. I was terrified that my life had stalled and that it would mean not a thing. So I entered into another phase that would turn out to inform my sensibilities in the digital world to come. I became an activist.

Inquests

The most terrifying thing in the 1970s was the prospect of nuclear war.

The nuclear arsenals of the Cold War were sacred, untouchable.

---

* This means of stopping a car might sound utterly terrifying, but in the early years of the automobile, it wasn't unusual, and in our impoverished region the early years hadn't quite ended yet. Not commonplace, but not outlandishly extraordinary.

The nuclear powers held each other in immobilizing choke holds. Ordinary citizens raged against a synecdoche: the more accessible targets of civilian nuclear electricity plants.

There were plenty of reasons to be upset with Atoms for Peace* in New Mexico. Relatively poor New Mexicans were being asked to subsidize a nuclear plant that would be built in relatively rich Arizona to serve very rich California. Meanwhile, nuclear waste was to be buried back in New Mexico, in an impoverished area near Carlsbad Caverns.

While I wasn't sure I was antinuclear in any general or absolute sense, there were plenty of specific, local antinuclear positions I could get behind. But one couldn't really fine-tune. What a beast politics is. One must excite paranoia and rage to get anything done and then hope they do not cause more harm than good.

I learned a little about law and hooked up with activists interested in the nuclear issue. After a while I figured out how to sue the big New Mexico utility over the subsidy issue. Hearings took place in Santa Fe, New Mexico's capital town.

Appearances were tricky for me. I had almost no money and overnighted for a few months in a sleeping bag under a bridge that covered the little creek by the capitol building. I kept a suit I could change into in a public restroom. I cut my hair, which responded by blowing up into a fluff ball.

I was able to get court-ordered access to financial documents and identified embarrassing fudging. It felt like a comedy writer was in charge of reality.

The utilities came up with outlandish ways to spend money, like crazy expensive custom party balloons, because they earned a percentage of what they could manage to spend.† What was most amazing was how the executives were incapable of seeing how ridiculous they had become. The experience taught me how power can make the powerful blind.

My stint as a pretend lawyer had a small impact, though nothing lasting. It was great fun to make fancy high-priced attorneys squirm, and amazing to confirm that an ordinary citizen really could have power in

---

* Atoms for Peace recalls both a famous United Nations speech by U.S. president Dwight D. Eisenhower and a policy of broadening the use of nuclear technology beyond weapons to manage the level of fear instilled by the dropping of nuclear bombs on Japan in World War II.

† This is an example of what is known to economists as the Averch-Johnson effect.

the system. I appreciated our country more. For a brief period I thought I might become a real lawyer.

One night I met a delicately dressed, well-kept hippie girl under the bridge. She was wrapped in pastel chiffon robes and had porcelain skin. We made out, starting something, and then her obviously wealthy dark, handsome boyfriend showed up and gave me the evil eye. Off they sped on his opulent motorcycle. Who are these strange creatures who walk among us? Today I would Google or Bing, but in those days one moved from mystery to mystery.

I went and got arrested in a carefully organized nonviolent demonstration at the nuclear plant site in Arizona. It was fun being in jail with a bunch of college types. One fellow I met there had made a practice of being arrested all over the world in nonviolent protests, including on Red Square in Moscow—no joke in those times. I admired him, though it wasn't clear if a protest of everything is a protest of anything. It was more a spiritual practice, and possibly a useful one.

While driving the Dart back to New Mexico, I picked up a hitchhiker, a young biker chick type, who believed in psychic phenomena, as many people did back then. Torn white T-shirt, leather pants, crystal pendant necklace, tiny nose, red hair, squeaky voice. She believed that people are always subconsciously trying to track each other, psychically, but most people aren't good at it. They need clues.

"You have to tell *no one* where you are going. Then, after a while, you feel it. It's amazing to shake off all the amateur psychic tendrils. People aren't poking you anymore. The quiet and tranquility are holy." I wonder where this woman is now and how she's coping with the Internet.

Back in New Mexico, I learned to stage media stunts. I had myself doused with water while holding a banner that read I AM DRY. This was to publicize the brine pockets found in the supposedly dry nuclear waste site near Carlsbad.

Consensus and Sensibility

An obsolete body of American law called the Fairness Doctrine declared that the public owned the airwaves that carried TV and radio signals. But, there could only be a few broadcast stations in practice, because that's the way analog electromagnetism works.

TV was powerful. It was becoming the essential platform in politics. If the owners of only a handful of TV stations coordinated to present biased or false news, there was no comparably powerful mechanism available for other parties to present an alternative.

TV station ownership might start to resemble a thought monopoly. Indeed, that's what happened in places like the Soviet Union. Why couldn't the same thing happen in the USA? If Democrats, say, owned all the TV stations in a region, then the Republican Party could just go home.

Therefore, according to the Fairness Doctrine, anyone using the public airwaves was obliged to show all points of view, not just preferred ones. TV was the public's resource.

It's an idea that sounds both radical and quaint today, and indeed it was snuffed out by Ronald Reagan long ago. But at the time, the Fairness Doctrine seemed sensible to most people across the political spectrum. Even so, the law hadn't been tested much. A batch of friends and I decided to give it a go.

We "intervened"—that's a formal legal term—when television stations in the El Paso area had to reapply for their rights to use the airwaves. This was usually a routine process, but we managed to tie them down in hearings and force the stations to fund a campaign of ads that would counter a previously overwhelming wave of pro-nuclear advertising paid for by the utility building the Arizona plant.

A batch of dusty hippies in New Mexico suddenly had a court-ordered budget to produce TV ads. How could we do it? The nonviolent movement was often the scene of social experiments, and one of these was "consensus decision making." It's like a wiki, except *everybody* has to be happy. There's no Wikipedian elite crew to shut people down. So the meetings take a really long time. You start daydreaming about what fun libertarians would be having by now.

We decided to create a television campaign this way. Hundreds of volunteers got together to deliberate on scripts, casting, locations; everything. Took months.

I didn't think the ads were all that great. By committee, after all. When they finally aired, however, a notable slice of the viewing audience had at least an indirect connection to someone involved in the productions, so our little ad campaign won outsized attention.

We had conjured that power that comes when masses of people feel a stake in a mass media expression. It's the same power that propels Twitter today.

After the thrill subsided, I began to have doubts. We didn't have much impact on immediate events. The plant was built, as was the waste site. We probably contributed a little to the slowdown of the nuclear power industry in the USA, but that wasn't really the desired outcome for me or for many of my comrades.

Persuading people to treat nuclear power as an evil makes no sense. It's just a type of technology. What should be happening is massive research.* Grassroots politics is too blunt an instrument to get this issue just right.

My adventures as an activist made me feel significant, which is what I desperately needed at that time in my life. But I realized that there was a hollow at the core.

There's currently no great way to integrate science and politics. I started to feel I was wasting my time, trying to stretch the tools of political struggle to influence what were really engineering decisions, which have a different character. (The same mismatch has since complicated climate change activism.)

There was also an interior problem with activism. You start to find your own worth in the cause, and that's too narrow a formulation. Activists start to fudge a little to reinforce each other. You pretend you're having more impact than you really are, and that you agree more than you really do. Some of my best friends in the "cause" became depressed from time to time, and a few committed suicide.

One morning I achieved clarity that it was time for me to move on. But to what? Love, it turned out.

---

* If engineers can demonstrate a design that is safe and efficient, irrelevant to weapons, and doesn't present a deadly, unsustainable waste problem, then nuclear power could be just great. It's an unknown if that will happen, but there's no proof that it can't be done.

7. Coast

Casual Enrollment

Cynthia. She played the cello. She dreamed of moving to Vienna, where my mother was from. Her parents played her Barber's Adagio for Strings every night as she went to sleep as a little girl. She was the first young woman I had ever met who I understood, at least a little.

She was in New Mexico to visit her mother, who had divorced her father long ago.

I can't help but write about this period like the silly young man I was. I didn't just feel attracted to Cynthia, but like the whole world was rearranged a little by her presence. The Organ Mountains, which loomed wonderfully above us, looking like a pipe organ, were no longer indifferent stone, but a stage set designed just for our frolics. Sappy enough? It's authentic.

She spoke about California as if that place was the heart of magic. The trees, the ocean. I had never gotten around to seeing the ocean when I was on the East Coast, so busy was I chasing urban dreams. I had only a hazy idea of what it would be like to stand on a beach. I painted a pearlescent ocean scene for her.

And then she was gone. Back to study the cello. In Los Angeles. I had to go.

Could the car make it to California? Could I afford the gas? A Buddha-like musician from California passed through town to visit an estranged girlfriend and offered to share expenses to get back to L.A.

The car with no floor had made long road trips before. I had taken it to Santa Fe, of course, and to Tucson for a science fiction convention. The Buddha and a gay couple, physicists, were my passengers when we were stopped at the Arizona border. A wax-museum-perfect cowboy-hatted patrolman in mirror shades poked his head in and asked, "Any fruits or nuts?" We all burst out laughing so hard that we were taken in for questioning. Eventually they had had enough of us and we went on.

This memory inserts itself because I can't quite make sense of it. I do remember having four people in the car, but there was no backseat. Did I improvise a backseat? On top of the hay? I can't quite put together a version of this story that makes sense. It goes to show what an artifact memory can be.

I got a ticket for driving on the highway through Phoenix too slowly. But once we made it between the preposterously high mountains around Palm Springs, the desert gave way to greenery and brown air.

Cynthia lived in a big classic Craftsman house in Pasadena. The poor Dart broke down practically the moment I pulled up. To my astonishment, Cynthia's family took me in, so there was no need to drive on.

She must have been cosmically shifted in time and space. Cynthia had a vaguely Central European accent even though she was raised in L.A., and a Renoir face, even though most of her friends were tanned beach figurines. Her cello playing sounded like the earliest recordings of classical music.

Pasadena was surreal, as this was not just a place, but the mysterious land of love. Everything about it was out of reach. Towering palms; impenetrable, mysterious air; manicured suburbs that stretched farther than you could imagine. I was told there were towering mountains right there, but the smog never cleared enough for me to see them. Recent immigrants from Africa walking in the land of cars, purchases from electronics stores balanced on their heads, looking as out of place as I was.

Cynthia, my obsession, turned out to be the daughter of the head of the physics department at Caltech. So that's where we would hang out. She was the darling of the marvelous minds who huddled there, like Richard Feynman and Murray Gell-Mann.

I was never a student at Caltech, but rather the weird boyfriend of the department head's charming daughter. It's a kind of status. Feynman was generous with me, showing me how to form geometrical designs with one's fingers to think about chirality; things like that. He was also a fun drummer and we'd play.

Oddly, Caltech didn't have much in the way of computer graphics going on at the time. I never found anyone there who shared my obsession with what a virtual world could mean.

But who cared, there was a girl. She showed me the ocean for the first time, on the way to Santa Barbara. The real ocean was brighter and more forceful than I had guessed, and smelled like life. I visited a little anemone in a tiny tide pool atop a boulder at my spot of first encounter for years, until a storm reshaped the beach and I could no longer find it.

City on a Pill

Cynthia decided I needed a tour of L.A. from a native's point of view. We drove her winged, pink 1960s convertible to Westwood on a Saturday night, and the street was filled with people dressed in candy-colored plastic outfits, pressing toward a pair of twin dwarves on the hood of a Cadillac, selling quaaludes.

We carried on for a few months; an enchanting, floating episode in my young life, but it was fragile. I didn't have a job, or any official status at Caltech. What was I doing? How could this last?

One day, disaster struck. I was dumped. For a pimply little physics student. She just told me, as if it was no big deal. We were just kids, after all.

World crashed. I had no idea what to do.

The bullet-riven Dart still wasn't fixed. I was broke and grounded, living in a house with someone else's girlfriend, haunted by the ghosts of limerence.

I needed a next step, so I started to explore the world beyond Caltech.

L.A. was a cypher. From the moment I arrived in New York City I had intuitions—that turned out to be mostly right—about the kinds of people who lived in a given building. But who were all the people who lived in the insular houses with the succulents and the driveways? No one I could imagine. L.A. never yielded to intuition, maybe because it was so enveloped by fantasies, both mine and everyone else's.

That's not all it was enveloped in. L.A. was polluted like New York, but with a distinct stench. New York was diesel, urine, cement and metal dusts from construction, moments of heavy perfume from people who walked by. L.A. was car exhausts. The toxic vapors of New York came from other people, but in L.A. they came from you. The back of your throat stung; it felt like millions of people being cooked in bad oil in a giant frying pan.

One day I had the very un-L.A. idea to take buses to visit the Watts Towers, and it took almost a whole day just to get close. I was walking along a street a few blocks from the towers when I was jumped from behind by four mustachioed white guys in gray windbreakers. They pinned me to the sidewalk and screamed incomprehensible orders right in my ears.

Then one of them said, "Hey, he's white!" They eased off me, and another said, "Say thank you."

"Thank you?"

"We're undercover police." Badges flashed. "Do you know where you are?"

"Watts?"

"This is a *black* neighborhood. Your life is at risk. You need to get out of here *now*."

"But everybody's been nice!"

"We just saved your life."

"Okay, wow, well, can you give me a ride to the bus stop?"

"No, you can't expect to freeload off the L.A. police." They hopped into a tan sedan practically in unison and were out of there before I could blink.

L.A. at large was depressing. Millions of people allowing their fantasy lives to turn their real lives into shit.

Cynthia's brother kindly offered me a way out: a ride on the back of his motorcycle up to Northern California, where I might clear my head.*

Rainbow's Gravity

I dismounted in Santa Cruz, a sparkling beach town with an amusement park by the ocean and a university up on a hill in a redwood forest.

Santa Cruz today doesn't feel as rainbow-strewn and romantic as it once did. It's always hard to know whether one's memories of how magical

---

* Just so you know, Cynthia and I are still friends, all these decades later. The connection was real. She's a professional cellist living in Vienna.

things used to be are just an illusion; what a gift to remember youth that way when you're middle-aged.

Even though I was broken-hearted, I was still in love, so my world was still organized around that precious sense that magic and meaning are nearby. Everyone was fascinating.

The magic I remember might have been more than the fog of love. This was back when silent springs were just a scary prediction. Insects, lizards, and birds accompanied you everywhere. Frogs roared in the night and huge native beetles would startle the squeamish in their sofas and beds.

California was more alive than it is today. Little frills of vines and moss poked through the cracks of even the most wretched stucco shack. The stars were brighter in the sky. You could see the Milky Way while lying on the beach at night.

Money was a stress. I shared a pathetic beach hut with five or six other teenagers, mostly students at the university. The rent was low, but still not free.

I made rent for a while by busking. I had my childhood plastic clarinet, a Bundy, and did reasonably well entertaining tourists for a few months.

Busking is the purest performance art. No one asked for you, so you have to win hearts based on nothing but the moment. I had my jokes and tricks, and learned to conjure a good attitude each day, a precious skill if ever there was one. Public speaking is easy once you've learned to work a sidewalk.

My anxiety about earning enough money to make rent, I later realized, served as a mask to insulate me from the more fundamental terror of mortality and the icy underlying loneliness that still haunted me from when my mother died. Capitalism gives us a faux death to avoid—destitution—and thus a ritual for asserting control over fragility and fate. It has its comforts.

Protogoogle

No such rationalization makes busking easy or reliable, so I finally had to look for regular work. There was certainly more of it than there had been in New Mexico. I answered an ad in the paper and showed up at yet another crappy stucco structure—a decommissioned hotel—that was nonetheless bathed in the amazing rainbow light of the coast, and gilded by the fuzz of ivy and wildflowers.

A young man, a real operator, met me. An archetypal junior shark of the kind that has become utterly ordinary, but that I had not encountered before.

While everyone else was still a hippie, he looked like what we'd later call a yuppie. A suit, a haircut, a status car. Young people just weren't supposed to look that way!

Inside a moldy workroom, a scruffy crew of hippie teenagers was tasked with calling strangers from long tables with telephones in order to sell a disorderly catalog of dubious merchandise. Magazine subscriptions, home inspections to assess vermin, and the like. It was a little like busking, but much easier. I made an astounding $119 on my first day.

Corruption festered at the core of the enterprise. Each day, our yuppie overlord would present us with an illicit list of phone numbers of people who were prime customers for whatever we had to sell. Some of our marks were about to make an offer on a house. They'd get the scare call about beetles and roaches. Others had just retired. They'd get offers for insurance or weird health products.

Each morning, the yuppie would show up with his little cache of phone numbers and decide which of us kids would get the most mileage out of them. Attractive female workers got the very best lists. "Tell me you're gonna turn this sugar into some cake," he'd swoon into a girl's ear, dangling a wilted, smudged page, torn from a spiral notebook.

He bragged about his cunning in getting the precious little lists, which in those days were handwritten. Bribes were involved, sometimes in the form of pot or LSD, and he often asked the pretty hippie girls from the phone crew to go with him to help seal deals. They'd meet people who worked at phone companies, police stations, or hospitals, often in alleys or parking garages.

I would have long ago forgotten this ridiculous job but in retrospect, I realize it was a microcosm of how Silicon Valley would come to function decades later. Whoever could get at the personal data would own the commerce—and politics and society. Data would become both the new money and the new power. I wonder what happened to that yuppie.

At first I was relieved to be making more money, but after a while the job made me sick from guilt. It was manipulative and creepy. Also repetitive and boring.

One day I asked the yuppie if he thought we were contributing to the world or just leeching off it. He looked me like I was a turd on his Mercedes. It was clear I was done.

"We're finding people who need things and getting them hooked up. Of course we're contributing!"

"But we're being paid to get at them before they can even see what their choices are. Aren't we screwing up the whole idea of a market?"

"Fuck you, man." Off I went.

Audience

I had earned enough to carve out breathing room, but I needed to find other options. It had not occurred to me that I could get work because I knew about computers. This might sound like an odd gap in my thinking, but it was before the rise of the myth of the dropout hacker from nowhere making it rich. Tech jobs were still mostly controlled by old-fashioned corporations or government agencies, or so I thought, and I had no degree; not even a high school diploma.

What finally prodded me to go over the mountains to visit Silicon Valley was not a job recruiter, but the weird hippie visionary lecture scene.

The California coast in those days was filled with even more self-proclaimed visionaries than it is today. It was easy to be invited to an event in an amazing home, perhaps a renovated old prospector's shack by a stream way up high in the redwood forests, to hear tales about how flying saucers, chanting, LSD, unconventional sex, or other exotica would save the soul and the world. A fair number of these events had a technological fetish, even so long ago.

The points of tech culture reference were different. Idealist techies might have been enchanted by Buckminster Fuller and his notion of world games, or of the Allende regime's lost cause to create a cyber-Marxist utopia in Chile.

This was the circuit in which I started to give talks.

I hadn't thought I was the kind of person who would enjoy speaking to an audience, but a public persona started to pop up within me like a baby desert plant that hides for years, only to reveal itself for the first time after a big rain.

My debut didn't go well, actually. I managed to get myself booked as one of the eccentric speakers at a "happening" in a converted barn up the coast. A batch of ever-so-clever graduate students from Stanford showed up, ready to fillet ridiculous hippies. They eviscerated me with hostile questions, and I was caught unprepared.

Busking had taught me how to entertain an audience, but I had never before experienced intellectual sadism. It hurt terribly, but I soon realized I was fortunate to hit bottom right at the start of my speaking career. After surviving that first night, what was left to be afraid of?

From about 1980 to 1992 I gave versions of my talk about VR thousands of times in every imaginable circumstance: tough Oakland high school classrooms filled with gang members, moonlighting prison guards looming at my side with raised baseball bats to discourage any trouble. Or the most prestigious gatherings of prime ministers and bankers in Switzerland, where we had to be flown in by helicopter and were watched over by stone-faced men in uniform wielding machine guns. Actually these two venues were not that dissimilar.

It was a matter of faith each time that my shy, awkward persona would get swapped out for a cherubic public speaker. This other version of me was confident and gathered every single person in view into a hypnotic rhythm of ideas. My model was Alan Watts. I have no idea how I did it.

My main task was to convey why the thought of VR—a crazy, extremist medium that would exist someday—made me happy. VR's deep mission, as I told it, was to find a new type of language, or really a new dimension of communication that would transcend language as we know it. That might sound like the most speculative, far-out plan, but the mission had a sense of perilous urgency for me. I believed it was necessary for the survival of the species.

VR was unnervingly hard to explain. There was no film of it, not even useful photographs. Certainly no live demo.

I'd start with an introduction to how VR would work; head-tracked rendering and so on. The topic was so exotic in the early days that people were visibly shocked on first hearing.

I still use some of the introductory ideas and images from my earliest talks. The spy submarine that surfaced in the first "about" chapter predates my first visit to Silicon Valley.

After the introduction, my talks dove into ideas about early childhood, cephalopod cognition, and how humanity would destroy itself unless art got more and more intense into the future, indefinitely.

Transcript

A transcript of one of my earliest talks has survived! And here is part of it, lightly edited:

Think back to your earliest memories, and ask the question, "What was I experiencing before then?"

There isn't any perfect way to answer that question. The answer hovers just out of reach. You can observe little kids, as Piaget did, even measure their brain waves, but the only way to get at what the experience might have been like for you is informed imagination.

Here's what I suspect happened before the times you can remember.

There was an early phase each of us went through when we were unclear about where imagination ended and reality began. It was a confusion that made us incompetent. If you can't tell whether a phantom is really there or not, it is quite difficult to navigate the world on your own.

During this period we were completely dependent on our parents for even the most basic elements of survival, not to mention comfort. But the internal experience of being so vulnerable was not negative at all; in fact it felt luminous, empowered, and even divine.

In that state, it was as if anything you imagined leapt into being. If you imagined a gem-encrusted tarantula emerging from an open window, it was as real as the window.*

If you can't tell what is real, everything is real. Everything is magic.

This is a way luckier identity than being King Midas. Everything he touched turned to gold, but anything you merely imagine turns real. You are a god.

Then a horrendous tragedy overtakes you. You finally start to distin-

---

* Why a tarantula? I had just hiked a mountain in the Bay Area where they swarm to mate.

guish what is real from what is only imagined. The window is always there but the sparkling tarantula sometimes is not. Other people acknowledge the window but not the creature. The window and the tarantula are not of the same world.

This realization grows into a belief in the physical world. The physical world is the one in which your body exists and you learn to control it. In time you will walk, run, and speak.

But the realization also amounts to a severe insult. It is the most precipitous demotion possible in any possible world. One moment you are Lord of the Universe, willing things into being, and the next you are a small, damp, pink thing, helpless in every way.

It is a bitter pill to swallow. I suspect it has to do with the "Terrible Twos." You don't give up power willingly or gracefully. At every step you test the physical world, hoping to find a trick, a hidden angle, that will allow you to recover a little of the Protean abilities you lost so recently.

The struggle goes on for months, and even years and decades. Other bitter pills present themselves, such as the awareness of mortality. At the conclusion of your epic fall from grace, you are finally an adult.

Some people never quite make it.*

Most of us probably haven't quite accepted the transition with all our being.

Becoming an adult does not mean a total loss of creative power. It just means you have to put up with an enormous amount of inconvenience.

As a child you might conjure an amethyst octopus friend, 200 feet tall at its center, with tentacles 400 feet long. It wanders into town when you call it, and sleeps under the waters of the Bay at other times.†

The octopus bends down to give you access to the top of its head, where there is an opening. Inside the head is a wonderful, furry cave where you can hang out. There's a little bed in there that hugs you while you sleep. Imagine this at bedtime, and it is as if the creature is real.

How long does it take to realize a dream? A young kid can imagine

---

* Emphasis on "some"; looked around at the audience accusingly.

† There's also a story behind this choice of creature. Around the same time I would have given this talk, a few friends and I tried to launch a robotic Loch Ness–type creature to inhabit the opaque waters of the San Francisco Bay. It was intended to be undetectable most of the time but would rear up into the air near tourist areas like Fisherman's Wharf once in a rare while.

the mega-octopus with the sleep chamber into existence in, let's say, a few seconds.

As an adult you can imagine this creature, but that alone does not make it real. Something is real only when it is also experienced by other people —and not just as a movie, but as a world to be explored. As a world that anyone might change, because shared consequence —shared experience of change —is what makes a world real.

The previously available reality-based option was to use technology to actually craft the creature. A giant robot? A genetically engineered giant octopus?

Before virtual reality, making a fantastic scenario real not just for you but for others was sometimes possible, but meant time and hassle. A gigantic, catastrophic wall of hassle! Life isn't long enough.

Virtual reality tugs at the soul because it answers the cries of childhood.

There's more, and I hope you'll read it. The rest of the early talk is found in appendix 1.

Giving talks became an ever-present stream in my life, even later on when I was out of my mind from sleep deprivation, running a tech startup. As you read the rest of this book, keep in mind that through it all, I'd still find an occasion every couple of weeks or so to speak about VR and the future.

The anthropology of the visionary social circuit fascinated me. In fact, it is what motivated me to make it over the mountain pass to Silicon Valley.

I noted that techie oddball hippies were rich, while all the others, except the drug dealers, were poor. I finally got a clue.

8. Valley of Unearthly Delights

Now we come to my story of Silicon Valley in the 1980s. I can tell the broad outline of what happened in one long sentence: I started a career in the young video game industry, made money, used the money to finance experiments in what I called virtual reality, met kindred spirits, started the first company to sell VR gear and software, prototyped the major applications of VR such as surgical simulation, helped create a cultural storm that swept me up—a psychedelic party and publicity scene that celebrated VR—and then left for New York City after an almost surreal struggle over control of my company and other bizarre battles.

The texture of life changed. Beforehand I was a weightless rolling stone. When you're a massless particle, you are light; the world exposes impressions as you flash through.

Exotic road stories are fun to tell and fun to hear, but they depend on not having had a deep stake. When you alight in a place, you must deal with people for real.

When you're planted, you must also deal with yourself.

El Paso del Cyber

I hitchhiked down to L.A. to retrieve and repair the decrepit Dart, clinging to the coast on the way back, avoiding steep inclines. My next anxiety was whether it could make it over the mountain to get to Silicon Valley. It was not generally able to climb hills—certainly not at highway speeds.

One day I just had to try; poured extra oil in and started the ascent of Highway 17.

I was expecting an enchanted place on the other side, a techie version of Santa Cruz; *The Garden of Earthly Delights* but with blinking lights and whirring tape drives.

Instead, I found what looked like the most dispiriting parts of L.A. Freeways gnarled by sullen, low industrial buildings that were born ugly. It was in these lifeless places that Silicon Valley reinvented the world. Has there ever before been such an unaesthetic center of power and influence?

There wasn't yet a way to print from a computer unless you were at a fancy lab. But my father's old portable Royal typewriter had made it with me, stuck in the muck in the trunk of the Dart, and on it I composed a bare-boned tech résumé.

On reflection, I actually had done a fair amount. Research on an NSF grant, programming on a variety of computers, lots of math.

I parked my toxic jalopy around the corner, out of sight, and stepped into the only "headhunting" office I would ever visit; the blandest room on earth.

I still remember how I noticed myself at that moment. Instead of dropping off into a glazed trance while taking in the excruciating blandness, I caught myself and stayed alert. It was a foray into self-control; an attempt to live beyond the pervasive veil of "mood."

A woman in her thirties was working the front desk, makeup a bit overdone, skin oddly tight on her face, tense, a little angry, a little sad. She was dressed in the awkward women's business attire of the time—including a starched yet puffy bow that was supposed to reflect a man's tie.

"It's incredible when you see how it happens. Not to the people you'd expect." A reverent sigh as she sorted through papers.

What? What was she talking about? Winning the Nobel? Beatification? Of course it was about who got rich. Apparently a lot of people were paying attention to that, and finding it disconcerting that the new wealth seemed random. "Look at this one, just an ordinary engineer, happened to sign up with that stupid company, didn't have any ideas, did nothing." Ouch. Jealousy. Poison.

She brought me back to a cheaply wood-paneled room to see the actual headhunter, only a few years older than me. Suit and tie, fresh shaven, cold green eyes. He looked me over greedily, as if I were one of those lists of phone numbers of prime targets for the yuppie scammer.

"Can you start today?" What?

This was before computers were connected or could display much text; and as I said, there were no printers. (Years later, when it became possible for a few people to own printers, I used to joke that printers were replacing hot tubs as bait.)

So the gleaming fellow leafed through a crumpled, handwritten notebook, careful to angle it away so that I couldn't peek. The salaries he relayed to me—in firm, conspiratorial whispers, like a pusher on a street corner—sounded surreal, incomprehensible. I was disoriented and didn't know what to do. Could I really live and work in this place away from the rainbows, this anti-Narnia?

Optimal Us, and Them

The very next word I heard in Silicon Valley after my meeting with the headhunter was, unsurprisingly, "Hi," but after that came nothing but surprises. "The first thing you need to know is that there are two main tribes, the hackers and the suits. Don't trust the suits."

A friend of a friend from Santa Cruz gave me this advice. Hippie, careless exterior; rough fringed poncho, big black glasses; explosive beard like dark smoke. We drank smoothies at a natural food place near Stanford; outside table, sunny hot day, wood chips underfoot, girls in tie-dye glimpsed for a moment at a corner table, then gone.

"Don't get me wrong, we need the suits, but you gotta watch them."

Yet again, people were forming themselves into tribes, for no other reason than to find mutual mistrust.

"Suits don't really do anything except get paid to do the stuff that's so boring that no smart person could ever stand it."

I thought of the yuppie from Santa Cruz. Was it possible that there were other people like him? Whole hordes of them? Yikes.

"Suits are like women. You need to deal with them for there to be a future at all, but oh man what a pain in the butt."

I felt distress surging from deep within, nausea. What was going on? The ability to notice my own reaction was still fresh and uncertain. I strained to get a read.

Suddenly I got it. The women of the world were the shards through which I could hope to find a continuation of my mother. It wasn't well thought out or clear in my mind, but I vaguely thought of living women as being channels to my missing mother. I wanted to be in a place where a sense of her might come through. I had thought California wasn't so much a macho man's world as New Mexico or New York. That was actually so in Santa Cruz, at least once in a while.

What if Silicon Valley, the place I might have the best shot at earning a living, turned out to be a place that pulled me away from the female world, from any hope of catching hold of a trace of who my mother had been?

Flustered, I managed to say, "Are all the suits so bad? I have a friend who works for Steve Jobs at Apple and seems to think he has good ideas."

"Oh yeah, I worked with Steve at Atari, where he tried to be an engineer. The guy used to brag about how he'd optimize this chip, but I never saw him get anywhere. At least he learned his place."

What a curious society. Status was attached to technical attainment more than to money. (If "hacker" originally meant a person who was too smart to endure the boredom of dealing with the money stuff, then there are vastly fewer hackers in today's Silicon Valley.)

There was another term, "cracker," for those who broke into computers, but since computers weren't networked yet, there wasn't all that much cracking going on.* The hacker/cracker distinction was not between good and evil but between those who were better at creation or destruction. Destruction was thought of as a good cause for the most part, as the world we were given was so . . . what was the problem? The world was not optimized.

There was an awkward but passionate cowboy metaphor in play. We hackers were the roving gunslingers. We supposedly lived by a code. Ethical hackers and crackers were called "white hats" and the bad ones "black hats."

I had grown up around real cowboys. Some of them were kind; some were thugs, just like people anywhere. As a rule, a cowboy wasn't any more free than any other variety of person. So the hacker mystique was broken for me from the start.

Like cowboys, hackers supposedly experienced freedom in a wild land that yielded only to our special prowess and expertise. We roamed as we

---

* Decades later, when the computers had long been connected, "cracker" showed up again as a derogatory term for "white male who doesn't appreciate his innate privileges." Almost all the crackers back in the 1980s were crackers.

pleased, inventing reality for everyone else. Normal people would wait helplessly as we blazed their new world.

What surprised me over the next few decades is that all those alien, "normal" people, all over the globe, chose to buy into our myth. You let us reinvent your world! I'm still curious why.

Finite and Infinite Games

I went on job interviews for only a couple of days before choosing my first Silicon Valley gig. It's worth remembering those days of misstep, because first impressions can be so revealing, both of you and what you encounter.

I was groping toward a career in VR, but there were no jobs related to VR, as there was not yet a VR company. (And you couldn't just raise money out of thin air for a startup in those days.) No one even knew the term "VR." I couldn't hope to apply to places that worked with flight simulators, like NASA or the air force, without so much as a high school diploma to my name.

The plausible jobs that resonated best were in the infant video games industry, despite the way it repelled me. At least there was a semblance of art and music.

Repelled? Indeed. I don't like fixed rules. I loathe feeling like a rat in B. F. Skinner's lab, being trained to run again and again, ever more perfectly, on a petty course invented by a remote master. It's even more disturbing to imagine hundreds of thousands of people simultaneously running a maze that I might invent.

Loads of people in the technical world are obsessed with the kinds of games that I find boring, and humiliating, in a way, because of the lab rat role you must embrace. I see these games as math's way of portraying moral and societal failure.* Life should be about rejecting claustrophobic games like that, not becoming adept at them. The most important math concerns avoiding games with fixed rules and sharply defined winners and losers.

---

* The Prisoner's Dilemma is one of the most famous game theory thought experiments. It has been adapted into game shows and movie plots. I won't explain it here; look it up. It's interesting from a mathematical point of view, but it's a horrible way to think about real life, which is never so clear cut. I find it heartbreaking to watch people on game shows or other real life enactments of the Prisoner's Dilemma learn to become cruel and deceitful to one another. I suspect the stench of this use of math has turned off as many kids who would otherwise like math as the usual demons like awful teachers and textbooks.

And yet, games were the only interactive art form that was making money. How could I not go there?

My first job interview was actually all the way up in pretty Marin County, over the Golden Gate Bridge. George Lucas was starting an organization to create digital effects for movies but also for video and audio editing, with ambitions to get into video games. You might think I was interested because of *Star Wars*, but no, it was because a student of my hero Ivan Sutherland, a fellow named Ed Catmull, had started these digital efforts.

I entered a big, unmarked industrial building and was greeted by a giant painting of the Organ Mountains, the pinnacles I stared at so often as a child in New Mexico. How could this be? Turned out that one of the other prime digital gurus at the place was Alvy Ray Smith, another migrant from our corner of the desert.

It was cool but a little disorienting to see Alvy, as if universes were colliding. He had grown up just a stone's throw from the dome. I knew him mostly for his wonderful work extending something called Conway's Game of Life.

The Game of Life was a program—created by mathematician John Horton Conway—that showed a grid of dots that blinked on or off according to simple rules about whether neighboring dots were on or off. By tweaking the rules and the initial pattern of dots, you could get amazing, unpredictable things to happen, as if the Game was a miniature living universe.

Alvy proved that you could even cause a fully powered computer to come into existence within the confines of the game, so that there could be worlds within worlds—an insight that would be popularized by Stephen Wolfram years later. It was natural to speculate that we might be living within something like a Game of Life.

Here was a "game" that expanded. It didn't stick the player into a tiny abstract prison.

Alvy's work comforted me. Once I understood that a deterministic game like the Game of Life could produce unforeseeable results, then a dark anxiety I had felt melted away. There was no longer tension between determinism and free will. If the only way to know the future was to actually run the universe, then it no longer mattered to my philosophy if there was a deterministic floor to the thing. There might or might not be. We could never know, from inside the universe. Moot point.

Of course the most *useful* physics might include randomness, or not, but that no longer mattered to philosophy. Math doesn't kill freedom! Faith in the reality of free will makes as much sense as its rejection.

Hackers used to argue about these ideas all the time. "The ability to reject the idea of free will is an example of free will." "Are you saying that thing you just said couldn't be stated in a universe where there is no free will? Wrong! I can write a program that says it right now."

Alvy as a person is as comforting as his math. He has a cheerful approach to computers and life that I still enjoy. Abstractions are emotional! A physicist who works on theories in which the universe is emerging and unpredictable will tend to be warmhearted and funny, like Lee Smolin, for instance.

But let's get back to my story.

Loop Skywalker

The person who interviewed me wasn't Alvy, alas, but yet another polished young Suit. He obviously wished he were working in the glamorous movies instead of in what were still considered the second-rate digital boondocks.

"We want to eventually bring *Star Wars* to life, so that it's *you* directing Luke Skywalker. You'll be able to make him swing the light saber with the joystick. Do you think you could make a digital light saber look like it's glowing on one of these 8-bit machines?"

"Oh, you know, I don't think this is the job for me."

"Wha . . . How can you say that? This is the biggest thing ever."

"I don't mean any offense. It should be an amazing job for the right person. I just don't like *Star Wars* that much."

"What the fuck! Why are you here?"

"Well, I didn't know what the job would be."

"How can you not like *Star Wars*? Everyone loves *Star Wars*!"

"Oh gosh, I don't hate it . . . I can explain if you want."

"Oh yeah, I have to hear this."

"Well, look. A few years ago, when I was a kid, I used to play music to accompany Robert Bly when he gave readings back in New Mexico."

"Who that?"

"The poet? You know . . . He was reading his translations of Rumi—the ancient Sufi." I was clearly not communicating. "Uh, that's the sort of hippie mystic part of Islam—goes right back to the origins—*anyway* . . . We got booked along with Joseph Campbell, who would give talks."

"Oh yeah, we all know about him. George used his book *Hero with a Thousand Faces* as a template for *Star Wars*." As if this guy was best buddies with Lucas. "Wait, you *know* Campbell?"

"Not really, just performed on the same billing at this retreat place by some hot springs."

"I don't believe you."

"Well, yeah. Anyway, Campbell—really great guy—has a theory I don't really like much, that all human stories are variations on the same shared story. Kind of like how Noam Chomsky says there's a core of language."

"Don't know this Chomsky, but hell yes, if you make a pure version of that universal story, you strike gold. We did and keep on doing it. What's your problem? You hate money?"

"It's so confining. Not money; I mean this idea about stories. What if we don't really understand the stories of other cultures? Who are we to say they're telling the same story we are? And if there's only one story, how can we have hope that stories will get better in the future? If we believe there's only one story, maybe we're trapping ourselves in a small loop, like we're in a primitive, crappy computer program. Alvy, who works here, proved there can be these expansive kinds of programs . . ."

"What the fuck are you talking about? *Star Wars* is set in the distant past, not the future, and it's cool! Robots, faster than light spaceships! That would be a great future!"

"But the people are the same. They're stuck in stupid petty power games. They're cruel and selfish. Even the good guys are clannish and macho. Who needs more royal families? America was all about getting rid of them."

"Oh god, you hippie idealists are so full of it."

"Oh, don't say I'm like that—am not! Um, but, science fiction can be about people getting better, not just gadgets getting better. I mean, in *2001: A Space Odyssey* there's this sense of transcendence, like we might outgrow our petty little conflicts. Well, maybe that's not a great example—it's pretty abstract and amoral. What about *Star Trek*? Gene Roddenberry had this idea that people would get kinder as the machines got better. That's so much more exciting. I think it's actually happened in human history already."

"*Star* fucking *Trek*?"

"I guess I should go. Would you say 'bye to Alvy for me?"

"Not a chance."

You Have to Get Awfully Weird to Avoid Becoming a Behaviorist

So, the Lucas world was not my cup of tea. But what an amazing feeling to be able to say no to an opportunity like that. I could take my pick of hundreds of jobs.

It was the 8-bit era. I coded a few games for various companies and was paid insanely well. College flunk-out debt gone like a dandelion in the wind.

It was especially satisfying to design sound effects and music. In those days the programmer might do everything from art and music to the instruction manual.

I wasn't the only immigrant to Silicon Valley thinking this way. I started to meet other game hackers who thought of themselves as artists and scientists, and some of them would eventually help found the first VR company, VPL Research.

I met Steve Bryson, hippie physicist musician, dressed a little like Robin Hood, when we were both coding 8-bit games in a generic low-slung Sunnyvale office building. It had the usual grooved, prefab cement exterior; perfunctory hedges around the parking lot; show-off cars parked by the front door, my Dart parked around back.

A sterile habitat hosted marvelous, exotic people. What amazes me most when I think back on those days is how many incredible coders were also accomplished musicians. I remember storming into piano stores with five or six buddies, and everyone could not only play classical pieces expertly, but each was literate in jazz and had a personal, refined style. Steve Bryson, David Levitt, Bill Alessi, Gordy Kotik.

By 1981 I finally codesigned my first commercial video game. I worked with a toy and game expert named Bernie DeCoven on a title we called Alien Garden. It did fairly well. Then I designed my own game for the first time.

This was Moondust, which was a Top Ten home computer game when it was finally released in 1983. (The world was slow back then, before it was optimized. It could take years to release a program.)

Moondust was sold in boxes! You used to go to a big store that mostly sold vinyl records

Steve Bryson.

and there would be a section with video game cartridges. I was infinitely proud to see Moondust on sale in its custom floor display, with promotional posters high up on the wall.

The best version ran on a Commodore 64, if anyone wants to dig it up. The music was algorithmic and pretty, with an echo and wetness, which was a trick in those days. The music was driven by the action, a first in gaming. The graphics had a glimmering soft quality instead of being blocky. Also quite a trick back when silicon was slow.

But the gameplay was bizarre. You influenced a whole swarm of spaceships at once to try to get them to smear a flowing ribbon of color into a ghostly, shimmering target, which would undulate orgasmically if you succeeded. The gameplay was too complex to approach analytically. One had to become intuitive, and then there was the oddly sexual quality.

It was amazing that throngs of customers bought copies of Moondust. I suspect they were drawn in by the graphics and sound but then quickly gave up on playing it. It was too strange, too open.

Grounded

Shortly after I arrived, I had rented an uninsulated hut in Palo Alto; old railroad worker's shack on a dirt road, tilted on borrowed time in an undevoured scrap of orchard by a creek.

You could tell who really understood Silicon Valley by their attitude toward real estate. I remember the real estate agent saying something that revealed her to be one of those horrid creatures who didn't get it.

"You're crazy not to buy a Victorian bungalow; those will be worth ten times as much in a few years."

One of my hacker acquaintances was nearby and set her straight. "Code will run the world directly. Money is only an approximation of the code of the future, while we wait for computers to get cheap and connected. We're creating a kind of power here that is much more important than money. Money's obsolete, or will be any minute." Yes, hackers talked like that. Everyone felt the call of oration.

The real estate agent looked to us like an uncomprehending dinosaur, staring right at the asteroid of doom.

My old home, even the ancient creek, has been untraceable for three decades now; instead, a mapping satellite detects only a smear of self-similar condos. I remember the rural smells of the gravel in the road and the moldy stained wood, the same inside and outside. California used to smell like grass, and it sounded like bugs and frogs.

Palo Alto was the spiritual center of Silicon Valley; less dismal than the grim empires of blandness down the road, like Sunnyvale, but still too dismal for me.

I stared up at the tall trees every evening in the endlessly perfect weather. The sky was always empty. No distant desert vista, no infinite sea, not even the wretched-but-fascinating urban New York mulch stretching to the horizon. All we could know was garden paradise, as imagined by immigrants from the snowy reaches of earlier American wealth. As if demons had tricked us with a simulation of heaven. Such a limited place, so discordant with my interior.

I felt dead lonely inside for years, for decades after my mother's death.

A Club That Would Have Me as a Member

Hackers were always showing off their latest projects. Since the computers weren't connected, you'd have to drive to see a demo, or bring it along with you. Instead of goats, the back of the Dodge Dart was now piled with computers, so I could haul around demos of my work. I remember having to occasionally pick bits of old hay out of hard drive slots.

I'd show everyone Moondust. I showed it to Alan Kay and his team up at Xerox PARC, to the people at Apple who would end up creating the Mac, to Doug Engelbart's group at the Stanford Research Institute, to the people at NASA working with flight simulators.

One time I hefted one of those giant old CRT monitors up onto a table at this dimly lit dim sum place in an alley near Stanford—to show people Moondust, of course. (Don't recall the name of the joint, but if anyone wants to figure it out, this was the one that put almond oil in their shrimp dumplings, and everyone talked about that.)

The diners that day would go on to found companies like Pixar and Sun. Moondust was a hit with this crew, and they started pestering me.

"How'd you do it? There are pixels changing all over the screen at once."

"Oh, I'm using a compressed lookup table through these shifting masks . . ."

"Wait! Don't tell them how you're doing it!"

"I thought the hacker ethic was all about sharing code."

"Well, yeah, if it helps bring down the big, bad old power. But this is your personal stuff."

"I don't know what to do."

"Well, anyway, you're one of us now." *One of us* said with that emphatic, grunting rhythm from the movie *Freaks*.

Code Culture

Our world wasn't made for us, yet. We were still profoundly strange.

The Valley already had elite pockets, but it was mostly not so rich, and much of it was raunchy and depressing. All of America, including the Valley, retained a slimy coating from the 1970s. Rusty signs with missing blinking lights offered live sex shows just north of Menlo Park, and beleaguered streetwalkers crowded the corners.

And yet this was our gathering place. We needed to stay close together, as there was not yet an Internet, but we needed network effects.

I remember playing pool at a rough dive bar on El Camino Real, the main drag, and thinking that a hacker in Palo Alto was like a cue ball that spins in a fixed spot after knocking another ball into faraway action. We spun in place in our new home while our momentum was transferred outward, reformatting the whole rest of the world.

Coding all night long, all the next day, coding until your brain absorbed a big abstract structure and perfected it. The experience was different than it is for coders today, because at that time you worked directly with the chip to get decent enough performance. That meant you weren't dealing with languages, tools, or libraries from other programmers.

Everything important was fresh, entirely made of your own mind. You were an abstract explorer, facing only wilderness. If you wanted to get a circle to appear on a computer screen, you had to figure out a way to code a circle that would be fast enough to matter. I remember going with Bill Atkinson, who coded the graphical aspects of the original Macintosh, to see the legendary guru of algorithms at Stanford, Don Knuth, to present new ways of drawing circles. It was like visiting the code pope.

Push anything far enough and it transforms. This principle applies even to computers. At the core of the coding experience, when you are functioning at the very highest level of excellence, you reencounter a mysterious sense of the world that is not code-like.

There is—or at least there used to be—an amazing feeling in the gut when code was correct. An incredible, almost messianic feeling. We used to talk about it with a bit of embarrassment, our hidden store of mysticism buried under a fortress of rationality.

Whenever I had that feeling, the code in question would thereafter prove to be bug free. It was a strange and almost holy moment, and you only got to feel it once in a blue moon.

That experience of the apex of programming has become ever more elusive, because programs are no longer ever written by a single person; new programs of any importance are usually made by teams, and when they run, they're spread out like moss grown upon myriads of preexisting software structures, which in turn aren't even running on an identifiable computer, but instead roam secretly between the world's uncharted, interconnected computers. One can no longer really know a piece of software; one can only test it, as if it were a newly discovered piece of nature. Yet another link to the old world of intuition severed.

Anyway, after days of concentration would come sleep like a velvet sea, often in your clothes, and then you might venture out and see other humans, but all of them had been doing the same thing. You looked like code to each other. You'd speak of the world as if it were an incomplete puzzle you were inventing.

I wish I remembered the names of all my earliest friends in the Valley. At least I remember the conversations. "I've kept data on all the sushi bars so that we can choose the optimal one." "So have I." "Did you timestamp your data? We could use a Bayesian method to correlate."

This means of encountering the world was still done on paper! We carried around mini-notebooks and pencils. Hackers would mount their notebooks in little pretend metal cases to simulate what it would be like to have portable digital devices someday. There were a lot of fancy belt mounts, wrist mounts, and vest mounts. After all that calculation we'd eat our sushi and then go back to our coding.

After you spend all day coding, you dream in code; you think of the world as code. Scott Rosenberg wrote a book that in part recounted my experience of dreaming in code, called, naturally, *Dreaming in Code*. You'd wake up and realize you were coding in your sleep, coding the events that were taking place around you in the dream. A loop for one's heartbeat.

9. Alien Encounters

The Essential Bug

It was great to find commonality with the other hackers, but I didn't really fit in. Most of them had different beliefs about the basics of reality and being a person. I was becoming more of an outlier with each passing day.

A new philosophy was rising and I didn't buy it. For me, the world was *not* code, at least in the sense of code we could ever know how to program. And people were *not* just more code, and the purpose of life was *not* to optimize reality.

My life was stretched to extremes by sleep deprivation and ambition. I became susceptible to extreme feelings. The new normal mind-set didn't just bother me; it made me want to scream.

Boy was I ready to argue. "The real world is a sea of mystery; we huddle on our tiny island illuminated by science and art. We don't know if the ocean has an edge. We don't know how much of it we see. We don't understand our place in it."

"You sound like someone from Marin."

"Do you mean that as an insult?"

"Yes."

"You're wrong, it's not like I'm from Marin. Those people believe in things without evidence. Like astrology."

"Um, maybe no one told you, but you can't have sex—ever—if you make fun of astrology around here."

"I know women who don't believe in astrology!"

"But are you having sex with them?"

"With one of them, yes."

"No way!"

"Look, I'm saying the exact opposite, that beliefs have to earn their ways."

"By that standard, astrology is earning its way for me. (Snort.)* How is consciousness any different from astrology? It's just this thing you believe in because you want to."

"I experience it. Do you?"

"What if I say I don't?"

"Then you'd be qualified to be some supernerdy philosophy professor somewhere. Go do that instead of programming."

"At least you didn't call me a 'premature mystery reducer' again. How do you know your supposed experience of consciousness isn't an illusion?"

"Consciousness is *precisely* the only thing that is still just as real if it's an illusion. Illusions rest on consciousness!"

"But then consciousness is not part of science. It's some kind of isolated singleton set that doesn't matter. Why even bother with it?"

"Admitting that a mystery is there is what makes us humble and honest. Without that, we couldn't have scientific method. Instead we would only make up code and then more code on top of that code. Our science is a genuine confrontation with mystery, and so is our art. There's mystery all over the place, every second. Reality is exactly the thing that can't be

---

* Hacker culture was more or less a subset of hippie culture during this period, and hippies often felt entitled.

There were hackers, for instance, who thought that sex was supposed to be "free" like software or air. Consider the slogan of a techie commune in San Francisco that we all used to visit: "Every human being deserves to have enough air, water, sex, food, and education." The implication at that time was modest—almost ascetic; only "enough" and not in excess, so that there would be enough to go around. Communitarian, sustainable sexual entitlement. Meaning, by the math, sexual obligation.

Why did I even bother to argue? "What if a woman, or a man, for that matter, doesn't want to have 'enough' sex from someone else's point of view?"

"You're worrying about a problem that doesn't exist. Everything balances out."

"But what if it doesn't?"

I'd eventually have a version of this argument with every imaginable California utopian. Libertarians, socialists, AI idealists. They all discount the possibility that someone might not fit into a "perfect" scheme, whether about sex or anything else.

measured, described, or replicated to perfection. Consciousness is a great way to notice that. Admitting it exists makes science stronger."

I can see now that I must have been an annoyance, spouting off like a professor when we should have been rushing through the sushi to get back to the only thing that really mattered, the code.

Rent-a-Mom

Hackers were typically horny young men, but also sweet, and often struggling to have relationships. Our sweetness was complicated by our fanaticism. We were utterly committed to a new way of understanding life that seemed to be emerging from tech culture, and that didn't make relationships easy.

"She wants me to share all the chores, but the chores are stupid. I mean, who cares if our clothes are ironed? In just a few years a robot will iron your clothes if you want, or we'll program DNA to make bacteria that grow new unwrinkled clothes every day, or something. Why should we make ourselves miserable when there are only a few years to wait before the problem goes away?"

When I first arrived, a trope that came up again and again was that someone had started a company called Rent-a-Mom. How did one find this company? It wasn't in the phone book and there was no Internet yet.*

I remember discussing this once, when a few of us furry hacker youths huddled at a table at our beloved tiny Hunan restaurant, a haunt for mathematicians. You'd see people like Paul Erdos, the legendary roaming mathematician, concentrating in the odd stale light seeping from neon signs in the window. Certain hackers would always show off by ordering in Chinese, though the waiter never appeared to be impressed. No matter.

"I know a guy who uses it. Absolutely not a sex thing. Different middle-aged women show up and do his laundry and shopping, choose his clothes, listen to him complain, bring him food late at night, all that kind of stuff. They'll give him rides when he's hacked so long that he can't drive. He says it's made him ten times as productive."

---

* A search for "rent-a-mom" in 2016 yielded a number of nanny, home care, and au pair services. None of these contemporary concerns have any connection to the legendary 1980s specter, so far as I know. In our age of text-based search, every name for everything is used by somebody for something.

From a tiny hacker, with glasses that were practically bigger than his head: "Okay, okay, how do we find this guy? Repeat, HOW, HOW, HOW?"

Rent-a-Mom felt like a real thing, since everyone talked about it, but no evidence, no actionable connection appeared. The enigma grew into an obsession.

I reacted with anger. These people hadn't lost a real mother. The very idea insulted the most central meaning in my life. My precious darkness. So I played the pissed-off person in a lot of the Rent-a-Mom conversations.

"Do your own fucking laundry. Or don't do it. None of us care about whether your laundry gets done at all. Your code will be just as good, or bad." Ouch.

"You don't get it." Same old ultimate put-down. "Rent-a-Mom does all the real world stuff so you can code. Imagine being freed."

"But we're wasting our lives just obsessing about Rent-a-Mom. How is that optimizing anything?"

"Exactly! Someday the computers will be connected and we'll carry around little computers connected by radio and you'll just speak into a microphone, 'Rent me a Mom!' and presto!"

While the technology didn't exist yet, visions and arguments were already formed.

Someone else would say, "But wait, why hire real moms? Wouldn't it make more sense to have artificial intelligences, you know, robots, do the mom stuff?"

"Well, yeah, but you don't get it. Human rental moms are here now, and the AI and robots might take time before they're ready."

"No, AI will be working in three years, tops." Remember, this conversation took place in the early 1980s.

"Okay fine, but the year doesn't matter. We'll use real moms who need extra cash, but we'll only do that for as long as we need to; it doesn't matter how long. If it's only three years, then great."

"But AI is going to be working so soon, why even bother?"

"Don't worry, it's only a contingency plan."

"But it really bothers me. I mean, AI is practically here."

"Well, look, we won't pay the real moms much."

"Better not!"

The Internet, as most people experience it now, was born at that moment, in that conversation.

"Whoever runs Rent-a-Mom when the computers get connected will run the world!"

"Yeah, so we better run it."

"But what if we don't?"

"We will."

Young Guru of Loneliness

Terrifying thought: I was living in a place, not on the road. I could have a steady girlfriend. A real relationship. I could start an adult life. Eek.

Stopping forced me into a difficult process of self-discovery I had avoided on the road. It took years.

All the hetero men of Silicon Valley complained that there were no women.* The imbalance was happily mirrored just an hour to the north in the city of San Francisco.

There, one heard every single, heterosexual woman complaining that all the available men were gay. Thus, as if we were trapped in an ancient Greek comedy, we men of Silicon Valley migrated periodically to find women.

Many of the brightest young women studied therapy of one stripe or another. It was, at the time, the best available approach to studying people. We studied machines, they studied people, all of us doomed to live out clichés. (Today, the cliché continues in an updated form, and holds that bright, technically inclined women study neuroscience.) There were legions of female therapists and therapists-in-training. My friends and I tried to engage, but the language of 1980s therapy was often challenging to us.

We talked about it. Remember, there were no connected devices, no social media; not even email, except in the biggest science institutions,

---

* We all wished there were more female hackers. The act of programming was a female invention, for the most part, but since the end of World War II, the profession had gone more and more male. There was one woman who programmed a hit arcade game called Centipede for the first video game company, Atari, and a scattering of others around the Valley.

In those days, the larger culture did the exclusion, not giving Silicon Valley a chance to show what its true colors might be: We deeply wished more women were coming out of math or computer science departments, but they weren't.

I recall this as a genuine feeling arising out of a sense of equity crossed with arrogance, for we thought being a hacker was the most glorious and important thing one could do.

which we rejected as being part of the old world we would overtake. All we could do was meet and talk.

A typical complaint from a hacker might be, "She wants me to express my feelings, but then she says my feelings aren't my feelings. She says frustration isn't a feeling, but anger and sadness are. I just don't understand what she wants."

I was also known to emit this complaint from time to time, but I mostly had a more specialized dating dysfunction.

I tended to date women who were older, often in their late thirties, while I was still barely into my twenties. Early on, I remember meeting a girlfriend at a hippie vegetarian café in the city.

Me, staring into space as she walked up.

"Pay attention, you! Earth to Jaron."

"Um."

"You didn't recognize me! Again!"

"Oh sorry."

"You do realize this is all about your mother."

"Oh god. Please."

"You totally forgot her. You don't look at her picture. You don't talk to your father about her. It makes me want to cry."

"I remember what's important. Please don't judge me. We're all finding our ways to get by. You don't really know what I've been through. Maybe it wasn't just like what they say in those therapy books. Can't we move on to another topic? Anything else."

"You need to face your feelings. Don't you realize how you're crippling yourself? You've made forgetting into your coping mechanism. You can't even recognize people. I mean, how can you live your life by not being in it? How can I date someone who doesn't even see me?"

"You're exaggerating. I see you. You're beautiful. You're brilliant. It's just that my mind wanders a lot. Takes me a few seconds to shift from my internal space to this world out here—just a few seconds! How bad is that—I mean it's almost nothing. You're the one who's judging me. Just relax. Maybe there's more of me to see, too. Maybe you're not seeing everything."

"You're so classic! This is how all the Silicon Valley men are."

"There, can't you see it? Now you're the one making me invisible. You're just seeing the stereotype in your head."

"Life is too short for this."

Gone was she. But not the pattern.

I always felt like the whole universe collapsed after a breakup.

Since every available woman was a therapist, I picked up the language of therapy.

I could expound on my own odd case; explaining to anyone who would listen that I was looking for a mother and that my quixotic obsession hopelessly skewed all my relationships. What I didn't say, because I didn't really know it yet, was that I hadn't yet gone through enough grief to find the glowing core of gratitude that she had been there at all.

Realization

Yes, it's true. I didn't recognize my girlfriend.

This is probably a good moment to confess a little about my cognitive quirks, which I only started to become aware of around this time. Maybe letting you in on my deficits of memory will earn your confidence in my strengths of memory. Hope so.

You, the reader, might be thinking by now that this book is a work of magical realism. Part of the action has taken place in Mexico, even. That might be what's going on here, but not by design.

Alas, what I am writing is the closest I can come to realism. My cognition is oddly ill-suited to the task of reconstructing meticulously accurate history. One reason is that I have a more than moderate case of prosopagnosia, or face blindness. I generally do not recognize people on sight.*

I have friends who are well-known actors and I have no idea when I am seeing them onscreen. That makes me an awful or wonderful person for an actor to know, depending on whether a performance is any good. Only the face-blind can watch a movie for real, free of the hypnotism of celebrity.

This is why I might identify a person in this story vaguely as "the widow" or "the dad." I'd rather convey the authentic incompleteness of my

---

* No one knows for sure, but about one in forty people are thought to have the condition, and many of those, like me, do not become aware of it for years. There are undoubtedly others who never become aware.

You can compensate by recognizing people in other ways. Individuals can be identified by where they are found, and who with; by quirks of how they move or through strategic chitchat, or by fashion and accessory choices (the popularity of tattoos has been helpful).

memory. The events happened, but the cast is uncredited. (Naturally there are a few instances in which I have intentionally disguised a living person.)

Being honest about the limits of one's own cognition is at least a start at wisdom, but no more than a start. I managed to remain unaware of my prosopagnosia well into my thirties. Once I realized what was going on, I could put a name on certain troubles, but they didn't go away.

I have come to not only accept, but value my prosopagnosia. I'm a great believer in cognitive diversity. Unusual minds discover important things that might otherwise go unnoticed. Because I could not recognize people on sight, I had to become more sensitive to what they did and how they fit into the world if I was to be able to recognize them at all.

I'm thought of as an intelligent person, but I'm not sure intelligence is a phenomenon with a singular magnitude. All human minds I've gotten to know well have turned out to be more amazing than I imagined at first. It's just that we're all tuned into our world in different ways. (Years later I would coinvent digital devices that could recognize faces for me, but I ultimately declined to use them. Trying to be "normal" is a fool's game.)

There's still more to confess. I also suffer from a peculiar quirk in semantic memory. I wasn't able to memorize the names of the months in order until my midthirties. I have become more normal with effort over time and can now recite them.

If it's hard to remember the months, imagine how hard it is to remember parties. I still worry that I will suddenly come upon a person who will gush about the wonderful, transformative time we had together at a conference, concert, or other gathering.

Don't I remember the Woodstock-like virtual reality event where I gave psychedelic VR demos in the 1980s and everyone was so electrified that they were indelibly imprinted with a glow that persists to this day? Or don't I remember when I spoke at a medical school about surgical simulation? Or when I talked to a young computer science graduate student at a conference?

As with face blindness, it can be horrible to fail to meet the sweet, modest expectations of a nice person who is right in front of you. I have inadvertently insulted people, and my protestations that it's me, not them, have probably just sounded strange and off-putting. I often wish I was better at lying.

Part of the problem is that there have been too many similar events.

Our world is awash with supposedly elite conferences, trade shows, parties, and ceremonies.

The grace that compensates for my defects in memory is that I can tell when I can't remember, and when I can. When a memory feels true to me, even if it's only partial, it always seems to check out. The feeling is like the gut feeling of code being bug free. There is a sensation of truth itself, deep inside.

If I'm not great at remembering events, faces, or sequences, how do I know my life?

I remember experiences in terms of ideas; how a story I lived through illuminated a deeper question. My experiences become allegories.

I recall minutiae about conversations with people from decades ago that turned out to be important to me. The eccentric heiress, the Mexican general: all colorful, but also characters in the allegories that make up my personal cosmos. I remember Richard Feynman teaching me to make a tetrahedron with my fingers, Steve Jobs demonstrating how to amass the mysterious quality we call power by humiliating a hardware engineer while I shriveled at the spectacle, Marvin Minsky showing me how to predict when a technology would become cheap and mature (he used genomics as an example).

As I hope has become clear by now, I also remember intimate subjectivities: moods and aesthetics.

These are the two poles of how I experience my world: the overwhelming indescribable flavor of what's right in front of me, and the ideas, the trussworks of thought.

I remember my life in a kaleidoscopic manner, or maybe it could be called cubist. So let us now return to a fractured, yet hopefully reliable, canvas.

Despite Myself

Maybe I'm being too hard on myself, but the way I remember it, I occasionally—and stupidly—mistook a date as a tryout for material to add to a future "guru talk." I was portentous, laser-focused, with piercing eye contact and hands aflutter, puppeteering abstractions.

"Each person must have a vastness inside, like Carlsbad Caverns. There are surely flavors, strange lights, imaginary things that language cannot express. Somehow, by some light, most people are probably geniuses inside."

"Okay, Jaron, this is amazing, but slow down. Can I talk for a minute?"

It took me years to learn to cede that minute. A shame, but youth takes time. More a shame because my rant was all about people being able to reach each other.

"I'm almost finished, just wait a minute and I promise you can talk. Really, this time. The most fundamental contact with reality is through math; it's the most universal touchstone."

A feeling was welling up in me that people would eventually become less isolated from one another through technology. (I was certainly not imagining today's commercial social media replete with spying algorithms that organize and optimize people for the benefit of giant server businesses.)

Her turn. "I guess I did okay in math before I switched to chemistry, but I can't say it was deeply meaningful. Can we talk about something else? Just for a while?"

"I know a lot of people find math to be alien and forbidding, but that perception might not reflect the deepest truth. Maybe the universe can produce more difference than we find even on Earth. How bizarre could an alien ever be? Maybe there are creatures somewhere made out of tiny knots in spacetime itself, creatures who have never known the stuff we think of as ordinary, like liquids and solids. They might not have noticed that ordinary stars exist. But even the most bizarre alien would know math."

"Jaron! There's an alien right in front of you, and this alien wants to connect. But maybe not with math just this one time."

"Oh yes, wow, cute alien here, but, but can I just finish this one thought? Otherwise it won't let go of my head."

"Oh, go ahead." Resigned, leaning away, but miraculously still attentive.

"Would aliens know the same math we do? This is an amazing, tricky question; the aliens might know different math that does not overlap with ours. But if they could become aware of our math, they would have to agree with our take on it. We would find a commonality if both sides tried. That's what makes math magical."

"Okay, that sounds right to me, but why couldn't math be a kiss?"

"Great question." Affirmative response, but not kissing. "Just, just one second more . . . There's this technology that could be created someday to turn your whole body, and the whole world, into anything, I call it virtual reality, and you could become a topological form and intertwine, and, and . . ."

Kissing.

Math Against Loneliness?

As I *tried* to explain to that girlfriend, I was burning inside with a peculiar utopian obsession.

I felt the world needed a tool for the spontaneous invention of new virtual worlds that would express the stuff of the mind that was otherwise impenetrable. If you could conjure just the right virtual world, it would open up souls and math and love.

Leave aside whether the idea is ultimately mad for the moment, please. Any notion of actually implementing it back at the start of the 1980s certainly was mad. But I tried.

First, I connected with the small community of people interested in "visual programming." This meant you'd control what computers did by manipulating images instead of strings of text.

Computers were still slow enough that you could feel them churning— the speed was still just barely within the grasp of human intuition. Programming was wonderfully concrete. Since you could visualize the inside of the machine in your imagination, it was easy to imagine that it could also be visualized in computer graphics.

Programming was an Eden. Today it's a crowded bureaucracy; code is all about reconciling what you want to do with endless layers of preexisting structures in the Cloud. (Are we in the middle panel of computer science's triptych?)

I was not the only person obsessed with the possibility that programming could be made more visual and intuitive. I had read about Scott Kim in *Gödel, Escher, Bach,* and I knew Warren Robinette from his wonderful video games.* We'd meet and work into the night, drawing sketches of how people might invent digital worlds to connect with one another.

---

* Scott Kim is known for his symmetrical calligraphy and a mathematical dance troupe as well as

One of my weird little projects was a purely sonic general purpose programming language with no connection to vision at all, operated entirely by singing.

An odd development was that by around 1982 I had money. Video game royalty checks! It would have been perverse and bizarre to put this extra money into real estate or stocks. The only imaginable use was to create the dream machines that I was living to try.

It was inconceivable to build VR in a garage at that early date, even with a nice budget. Even with an infinite budget, you couldn't quite yet buy a computer that could render a decent virtual world in real time.

But it was imaginable that one could self-fund research in experimental programming languages. So I did.

Appendix 2 explains what I was up to, and I hope you'll take the time to read about it. For the moment, all you need to know to follow the rest of the story is that I worked toward a kind of programming I called *phenotropic*.

---

his work in visual programming. He was featured in *Gödel, Escher, Bach*, the bestselling 1979 book by Douglas Hofstadter that brought a digital perspective on life and the universe to the general public for the first time. Warren Robinette created Rocky's Boots, one of the first "maker" video games, in which players constructed functioning visual programs on the screen of early 8-bit computers. Warren later joined the VR lab at the University of North Carolina, Chapel Hill.

10. The Feeling of Immersion

The group that will start the first VR company forms.

Woman as Social Stem Cell

Grace might just be a real thing in the universe. I recruited fellow travelers to help implement utterly mad designs. This still amazes me. People in those days were open to being drawn into fantastical schemes.

I can hardly express how much gratitude I feel now to those oddballs who were up for the ride. Remember Steve Bryson, from the Sunnyvale video game company? Sure, I'll go slave away in Jaron's hut on a bizarre experimental programming language with no reason or destination in sight.

Everyone who gets into the quest for a new kind of programming underestimates how hard it will be. Steve and I were soon overwhelmed. We needed to find more people; brilliant people who could support themselves elsewhere, who were game for a laborious, inchoate adventure of potentially cosmic implications. Where to find them?

Back at the Hunan restaurant: "Time to call the Grand Networking Females."

"Who?"

"He should call the GNF of the North."

"No, the South."

All I could manage was, "It sounds like we're in Oz."

"You noticed!"

I don't quite know the best way to convey this side of 1980s Silicon Valley, but women served as the impresarios of organic social networking before there was an Internet. Commercial headhunters were irrelevant to the real Silicon Valley; they were nothing but small-time scammers who preyed upon newbies like me.

The way the Valley really worked was that a tiny number of unofficial, supersocial, superpowered women connected everyone, creating companies, even whole technological movements. Histories of Silicon Valley always mention captains of industry like Steve Jobs, as they should, but you never see the names of the women who probably did as much to design the place.

Linda Stone, aka GNF of the North in the Little Hunan, later became a well-known executive at both Apple and Microsoft at different times, but she also had an outsized and intangible early influence on the evolution of Silicon Valley. A list of accomplishments doesn't really capture her role. Linda got Apple into making "content," in those days still distributed on compact disks (remember CD-ROMs?), and started Microsoft's early efforts in VR. But the invisible story is that a lot of the hackers who ended up at one company or another, or on one project or another, were prodded into place by her.

Coco Conn, based in L.A. ("of the South"), knew everyone and was probably at least half responsible for connecting the VR scene of the 1980s. Professionally, she worked with children in VR and organized VR events for the SIGGRAPH (Special Interest Group on Computer Graphics conference of the Association for Computing Machinery). Margaret Minsky, based in the MIT sphere back east, belongs on the list. Marie Spengler, one of the key figures at VALS (the Values and Lifestyle Program), the Stanford Research Institute unit that transformed twentieth-century marketing, was another example.*

I don't remember if anyone at the time talked about how GNFs might eventually be replaced by the AI robots that would replace Rent-a-Moms any minute, but that's exactly what Silicon Valley has tried to do. It's called social networking, and it doesn't work as well.

---

* You can see Marie interviewed in the documentary *Century of the Self*.

My first long-term girlfriend in Silicon Valley turned out to also be a lesser-known GNF. I won't name her since she isn't otherwise part of the public story, but she led me to many of my fellow travelers. She was, naturally, working toward a scandalous PhD at Stanford concerning male sexuality, and lived at the center of the seed of hacker culture. An original Sun employee and the first employee of Apple were her roommates at the time.

Through her I met Ann Lasko and Young Harvill. They had taught art at a hippie college in Washington State called Evergreen, and had come to town to get PhDs at Stanford. Ann was studying industrial design and Young was a painter and holographer in the fine art program. They were married, which was a shocking novelty in our circles, and had a couple of delightful, highly kinetic little kids.

Impossible Objects

At the time, children were no more than a theoretical, abstract concept to most of us single hackers. I remember plotting with the other guys that if we ever had kids, we'd put them in VR goggles right at birth. Then we'd swap in bigger headsets as our children grew up, but only while they slept so that our kids would only ever know VR. Our children would grow up in a four-dimensional world and become the best mathematicians ever.

Decades later, when I cut my daughter's cord, I flashed on that forgotten pact. Of course I had not brought an infant VR system to the birthing center. Oh well. Later, when Lilibell was eight or nine, I told her about it, and she was livid. "I could have been the first kid in four dimensions and you didn't let me?" She then demanded 4-D VR playthings and got good at manipulating hypercubes, so the moral of that story is that it's never too late.

If you are tempted, please know that there's a consensus in the VR research community that kids shouldn't get into VR before about age six, and some researchers recommend waiting until eight or nine. Give them a chance to develop basic motor skills and perception within the environment in which the human nervous system evolved, okay?

My daughter has as much access to VR experiences as any other kid on the planet, and she loves it. This charms me more than I can say. But she also loves bouncing on a trampoline just as much as playing with VR. I think she has it right. VR should be enjoyed as one of life's treats, but not as an alternative to life.

I've noticed that kids balance their use of VR in a healthier way than they do with videos or games. More research is needed before you should let me get away with claiming victory, but this is exactly what we all thought would happen back in the early days. TV and video games draw people into a zombielike trance, and kids, especially, get stuck in it, while VR is active and makes you tired after a while.

Triptych

Ann and Young were both skilled illustrators; Young and I collaborated on a triptych that depicted natural or "straight" reality (Young's work), mixed reality (Young's again), and full-on virtual reality (weird me).

In our drawings, the same couple is communing, each touching the other's face, in each frame. This was the concept image I would later use to sell investors on the first virtual reality company.

I arranged the drawings vertically, so that natural reality would be on top. It must always remain on top if we are to avoid confusing and manipulating ourselves into knots.

Why was mixed reality in the middle? At the time, mixed reality* was the less radical, halfway measure. The radical, transformative stuff would be full-on

---

* To learn more about the original and evolving meanings of the term "mixed reality," see the upcoming section called "Flags Planted," starting on page 237.

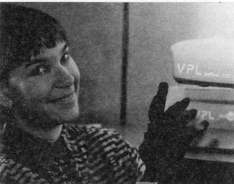

Young wore a strange outfit called a "suit" to demonstrate a DataGlove. Photograph by Ann Lasko.

Ann with a prototype VR headset.

virtual reality, the final frame. Mixed reality is harder to realize, however, so it only came into fruition decades later. Because of that, we now think of mixed reality as the more radical, more futuristic variant.

The old triptych is hanging in our home and still cheers me. It worships human connection. You can't see goggles, gloves, or any of the gear. It positions whimsy midway between straight reality and the impulse to make everything as exotic as possible all the time.

Maybe Ann should have run the company we would eventually start. She was the most emotionally mature of us all, our Wendy. She designed the first avatars and much of the early feel of virtual reality. Young was also a holographer and came up with new optical sensors for DataGloves as well as other interesting innovations over the years.

Ann and Young introduced me to other people from Evergreen, notably Chuck Blanchard. Chuck is one of the best programmers in history, confined to a wheelchair because of his MS. He was perhaps the sweetest and most pleasant of us all.

May I confess that it is terrifying to try to describe these people who meant so much to me, who were so patient and so generous toward me, in a mere book? How can I ever say enough?

Chuck was and remains a demon-level programmer. Fred Brooks had long ago observed, in his classic book *The Mythical Man-Month*, that programmers varied enormously in their powers. A single great programmer could often outrun a whole building of very good programmers. The best

Chuck, with dozens of DataGlove prototypes pinned to a wall in the background.
Photographs by Ann Lasko.

ones were legends. Bill Joy, Richard Stallman, Andy Hertzfeld* . . . Was I ever in their league? Maybe briefly, early on, when I programmed Moondust. But Chuck was beyond doubt one of the greats of all time.

Chuck had a genial lumberjack vibe, exuding casual but devastating brilliance. He was already in a wheelchair, but could still program with his hands. He had a neuroscientist girlfriend from Hawaii of such beauty that some of the other hackers used to slink out of the room when she was around because they found her mere looks intimidating. Not her fault.

Reality Engine Engine

The times were both electric and frustrating. I had vivid dreams and daydreams about what visual VR would be like when it matured. What would general-purpose VR headsets look like as a product? None were available for sale. Would they be held in place by elastic bands? How would you

---

* Bill was one of the founders of Sun Microsystems, which used to be one of the Silicon Valley giants. He wrote a famous cautionary essay about the future of technology called *Why the Future Doesn't Need Us*. Richard conceived the open-source movement, and you can read about how we argued in my book *You Are Not a Gadget*. Andy wrote the original Macintosh operating system.

make them? And what about audio? How fast would computers have to be to simulate 3-D effects?

We were living in the light before sunrise. There just wasn't yet a fast enough computer for the visual aspect of VR. Or at least not for VR that would be good enough to enjoy, much less apply to a practical purpose.

You might think that would have been a great moment, full of anticipation, a magic hour, but actually it was torture to wait for Moore's Law to climb a little higher. Like watching a teakettle that refuses to come to a boil.

According to the Law—and you were supposed to harass those who didn't "get it" until they ascended into our enlightenment—"Computation gets faster, cheaper, and more plentiful at an accelerating rate that human intuition finds hard to grasp." Those few who can imagine ahead can therefore seize the world. The Law was and remains the theology behind Silicon Valley's idea of destiny.

To those who "got it," Moore's Law meant we had discovered the ultimate possible ground-floor opportunity. Whatever we coded would inevitably reform the culture and politics of the world, the very fabric of human identity. This was not fantasy but rational extrapolation, and it has been borne out by events.

The sacred Law was intoned all the time. Our incantation sounded opaque to outsiders, but we knew its power.

It's still repeated all the time, even though it isn't so true anymore. Computers can't actually keep on getting faster and cheaper forever.

We can already glimpse a slowdown, presaging Moore's Law's Last Sigh. That might be like the traumatic American moment when there was no longer a frontier in the West ("commensurate with the human imagination"), to which the nation responded with a hollow Gilded Age. Not dissimilar to our situation today.

Allow me a metadigression to relate the argument about the Law that I always got into when I first arrived in the Valley.

Said I, "It's really only human understanding that improves; they call it an accelerating learning curve everywhere else."

That was not something to say if you wanted to win friends in the Valley.

"You don't get it. People are only the sexual organs that help machines reproduce and improve." Everyone was pseudoquoting McLuhan.

I shot back, "Look at what accelerates and what doesn't. Chips get better at an accelerating pace, but user interface design doesn't. The difference is that we define the edges and the function of a chip so precisely. Since we can nail it down, we can understand it better and better. User interfaces are about people, and people live out in the big unfenced world, and we can't specify them perfectly. So there's no way to achieve the same kind of learning curve."

"If you're right, we better find a way to constrain people more or the world will never get more efficient."

"Listen to yourself!" It was hard to argue with the god of optimization.

While I didn't think we should try to optimize people, I was all for optimizing computer hardware. I used to describe what the inevitable, sufficient VR computers—I called them Reality Engines—would be like. The 3-D calculations would be done in hardware. And not just for standardized objects like the terrain in flight simulators. A Reality Engine would be able to show any and all shapes, and as solid surfaces instead of wireframes. There might eventually be shading on a virtual object! A cube might look darker on the faces away from a virtual light! A concept so fantastic that the vision was hard to get across.

Should that be our little group's task, to make Reality Engines? No! A batch of Stanford people had just left to form a company, Silicon Graphics, to make exactly what we needed. But they hadn't gotten anywhere yet, so we waited and waited and we were getting antsy.

Much Touch

Even though it was too early to work with proper 3-D graphics, we could dive into haptics.*

Broadly, "haptics" means sensations that come from either sensor cells in the skin or in muscles or tendons—generally meaning sensations reported through the spinal cord instead of a dedicated nerve bundle between a sense organ and the brain. It is impossible to separate such sensations from human motion, so haptics isn't just about sensing. Haptics includes touch and feel, and how the body senses its own shape and

---

* Derived from the Greek *haptikos*, meaning "able to come into contact." It was proposed as a word in English in Isaac Barrow's 1683 *Lectiones Mathematicae*, but wasn't used much until recently.

motion, and the resistance of obstacles. It's surprisingly hard to define the term precisely because there are still mysteries about how the body senses itself and the world.

Haptics is at the very least how you feel that a surface is hot, rough, pliant, sharp, or shaking—and how you sense stubbing a toe or lifting a weight. It's a kiss, a cat on a lap, smooth sheets, and corduroy desert roads. It is the pleasure of the sex that made us all and the pains of the diseases that end us. It is the business end of violence.

It's the sense that overlaps the most with the other senses. We usually think first of the vestibular system of the inner ear as the body's means of sensing motion, but gravity and momentum are also felt by the whole body, which is, among other things, a tree of accelerometers. We sense sounds through our feet and indeed everywhere on the body, in the right conditions. (That has been what the term "club" has meant ever since the invention of the subwoofer.)

Does haptics include a stomachache? The boundary can be debated.

I love the haptic modality in part because we still haven't learned to work with it well or fully appreciate it. It's an intimate frontier.

We aren't as good at talking about haptics as we are at talking about colors, shapes, or sounds. What is the desirable feel of smartphone glass? Smooth, of course, but a sink is also smooth. It's cool-smooth, slidey-smooth, with just a bit of give, at the edge of sensation, and just a bit of grab so that you don't slide on it like ice. What is the word for that?*

We tend to use visual metaphors to convey analytic mastery, seeing a situation clearly, while haptic metaphors tend to convey intuition, gut feelings. Haptics is about you as a part of the world, not as an observer.

Other senses are aloof compared with haptics. The eyes and ears are interactive, in the sense that they probe by subtly and subconsciously shifting their positions (the "spy submarine" strategy), but haptics requires direct contact with the world. You push against things to feel them. You change them at least a little in order to perceive them.

Every touch carves away at least a shaving of what is touched. The stone in Mecca gets smaller every year, and faint gouges are being carved in the

---

* The people who engineer glass for smartphones are able to talk about these things using quantitative engineering terms, but their vocabulary has not bolted out into the language at large.

glass where you touch your phone the most. You are the weather of your world, slowly wearing it down. That is the price of sensation.

When I think back on it, it's funny how Moore's Law forced us to build useful equipment for each of the sensory modalities in a sequence instead of all at once; haptics first, then hearing, then vision, with olfaction, taste, and a debatable catalog of other senses still to come. Funny, because it mirrored the way I was slowly becoming more able to perceive the world as I emerged from the trauma of my mother's death.

Haptic Antics

Our haptics work blossomed under a member of the early crew who did not come to us via the GNFs.

Tom Zimmerman came up after one of my talks and said, "Hey, you talked about reaching into virtual reality. Guess what? I've built a sensor glove!" One of the wonders of Silicon Valley is that you just run into people like this.

Our meeting was fantastic. One of the huge impediments to our work had been that you could only interact spatially with a computer through a device like a mouse, light pen, or joystick—and mice were still exotic and hard to come by. However, if you could capture the whole hand's movement you could potentially pick up virtual objects, sculpt material, or even play a virtual musical instrument. A bridge between the way nature had made the human body expressive and the digital world was finally—literally—at hand.

Tom back in the day.

I introduced Tom to Steve, Ann, Young, Chuck, and the rest of what would eventually become the VPL crew, and we set to work. We built a series of amazing demos on 8-bit machines in my rickety hut. One of them was called GRASP, because it used a glove.

You could move your hand in front of the screen and a little computer graphics hand would move around with it, mimicking your gestures. Then you'd grab and manipulate shapes and forms in order to reprogram the experience as you were having it.

GRASP ran fast, since the images on the screen directly manipulated tight machine code—as I explain in the appendix on phenotropics. You could build games, mathematical models, and interesting art; nothing too complicated, but it was still crazy impressive for its time.

There was no business plan for this work, nor any academic glory sought. We didn't hope to publish or present at conferences. It was pure joy, to be shared over dim sum.

Alan Kay, arguably the principal visionary behind the way smart-phones and PCs work today, and a former student of Ivan Sutherland's, saw GRASP and said it was the best program ever written on a microprocessor (though one should be clear that his work at PARC was on a different kind of chip, called "bit slice"). Am not aware of even a video of that design that exists today. It is lost, like so much of early software culture that was built on old machines that can't be reconstructed.

11. To Don the New Everything
(About Haptics with a Little About Avatars)

Blind Bind

Some of the most popular VR headsets of the moment—around 2017—especially the ones based on smartphones—are bundled with barely any interaction. You just look around. Maybe you can press a button. How can anyone tolerate that? What is it with vision?

Vision has dominated culture for a long time. Visual records have transcended time and place since ancient times, and are the delineators of recorded history. To the degree that the sounds of language or music were recorded, it was until recently through visual notations. Actual sound has only been recorded for a century and a fraction, and haptics have only barely been recorded in early and extremely limited experiments. Until recently, vision connected generations.

Vision has inspired particularly self-conscious contemplation in the last century. Whole academic departments are devoted to distinct visual disciplines like cinematography, typography, photography, painting in its many guises, graphic design, and so on. We love to talk about what we see.

Vision makes us feel superior and invulnerable, like the eagle spotting a rat from hundreds of feet above. No one would think of placing an ear at the tip of a pyramid; it is the all-seeing eye that symbolizes the power of the dollar. That male habit of perception is called a gaze, not a sniff. You might have heard it through the grapevine, but you have to see it with your own eyes.

Can't imagine why I had a DataGlove in my mouth, but here I am in the long-gone cluster of rustic Palo Alto huts where we built VR prototypes before VPL came into being.

It might take generations before we fully internalize how obsolete visual dominance has become in the Information Age. Whoever has the best computer cloud will watch everyone else with superior intensity from now on. One's own eyes are increasingly beside the point.

Hand-Waving Demo; Digital Interface*

Tom, the gang, and I were obsessed with VR gloves before we made headsets and the rest of the gear in part because there was no choice. Computers were only fast enough for onscreen graphics, not in-goggle graphics.

But that was a blessing in disguise.

Input is more important than display. Your input in VR is you.

It mystified me to see people enthralled by the present-day fad for non-interactive VR experiences like the ones where you just look around inside a spherical video.

If you can't reach out and touch the virtual world and do something to it, you are a second-class citizen within it. Everything else there is connected into the fabric of whatever world it is, but you alone stand apart.

It's a subtle issue, best understood through personal experience; VR only exists because of subjectivities, but I will try to convey the feeling. To

---

* Hands, like butts, are uncommonly easy pun fodder; they're primal in human identity. If you meet me, don't go trying out a bunch of VR puns. I've heard 'em.

be an observer exclusively in VR is to be a phantom, a subordinate ghost who cannot even haunt.

Most people lose touch with the thrill of VR after the initial novelty if they can't interact and have an impact on the virtual world. Even the baby step of simply holding out your hand and seeing an avatar's hand that is still you, still responsive, still agile—this is a joy in itself. I never get tired of it.

In the following discussion of gloves, I'll speak of gloves as if VR headsets (as we know them today) were already working, too, which would actually not happen until a few years later.

With a DataGlove, you can pick up a virtual ball and throw it, or a virtual mallet, and then play a virtual marimba. It amazes me that anyone puts up with a VR experience in which you can't do these simple things.

Or, there's alien rock climbing. The cliff is alive and very tall. The handholds are constantly squirming around. Once you reach the top, which is unimaginably high, you can grab a hang glider and soar off.

People evolved with our hands! We ought to be able to use them.

Passive Haptics

Our earliest gloves were usually sensor-only, meaning entirely passive. They discerned the hand's shape but didn't directly convey any physical sensation. We experimented with various kinds of buzzers, heaters, and the like, but in truth none of these worked dramatically enough to become popular.

When fully immersed inside VR, users would occasionally report synesthetic sensations. I experienced them.

This is a trick I used to try at demos fairly often: Show a subject a virtual desk in a VR demo and ask the person to slam a hand down on that desk. If there's a nice shadow that converges with the hand as it nears the surface of the desk, and a satisfying thump at the moment of contact, most people's hands stop cold, even though there is no physical obstacle.

VR is like stage magic or hypnosis. Patter matters. If you confidently tell people that there *will* be physical resistance, then the hand stops even colder.

Fourteenth VR Definition: Magic tricks, as applied to
digital devices.

Naturally we dreamed of being able to sense what the hands and body were up to without having to put on special clothing. That finally became possible in the late 1990s through depth cameras. These are cameras that collect three-dimensional information from the world so that software can analyze what your body is doing. No gloves or body suits needed.

First-generation VPL DataGlove interfaced to a brand-new Mac in early 1984.

The image on page 132 is me as seen by an early depth camera (actually an array of cameras whose images were being compared in order to derive the 3-D shapes). This work was done in the 1990s when I was chief scientist of the Engineering Office of Internet2. It was a collaboration with a bunch of university labs, but the 3-D acquisition was largely due to work at the University of Pennsylvania by Ruzena Bajcsy and Kostas Daniilidis.

These days it's possible to estimate the shape of the hands or the whole body from extremely inexpensive consumer depth cameras like Kinect, first introduced for the Xbox gaming system by Microsoft in 2010, or the sensors embedded in a HoloLens. In theory, one should not need gloves ever again.

And yet, in my observations, wearing a glove lets the user's brain know that the hand is immersed, even if the glove is only a passive sensing device. The nervous system knows that something special is going on. Gloves might not be obsolete yet, even though I had expected them to be.

Bearing Arms

Gloves aren't perfect; there's a gauntlet of challenges.

Perhaps the most serious problem with the DataGlove was arm fatigue. Try holding your arm out without any support for a few minutes. You'll start to notice little tremors in your arm muscles and soon after that you'll wonder where all your strength went. We're used to either moving fast enough that momentum helps hold the arm up, or resting our hands at least a bit on the objects we're manipulating.

VPL DataGlove on the cover of *Scientific American*. Gloves were an icon of computing back in the 1980s.

A glove interface ended up with a starring role in the film *Minority Report*, and the production designers spent tremendous effort making that style of virtual interaction look sustainable when it actually would have caused arm cramps. I had brought a working glove-based surveillance system to the brainstorming sessions with the writers and Spielberg. It was a little like the one in the movie. Everyone could experience the cramps and understand how well the glove symbolized the way a computerized future could be uniquely seductive and uncomfortable at the same time. Ultimately, the use of the glove in the movie was deeply fitting: A real design that caused a strain was used to symbolize a fictitious future world that looked supercool on the surface but also hid its cramps.

Haptic Lemonade

In the early days, say up to about '86 or '87, we had to handle another glove problem, aside from arm fatigue. The computers weren't fast enough to keep up with human hand motions, which can be quite, um, dexterous.

Users would graciously slow their movements, usually without realizing

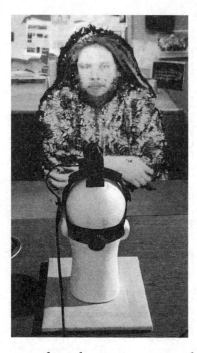

it, to give the sluggish sensors and computer graphics processors time to catch up. Glove wearers experienced time distortion: Users thought they were in the VR world for less time than they actually were. This was a cute demonstration of how the brain uses the rhythms of the body to measure the passage of time, but it was not what we were aiming for.

Our difficulties actually led to happy discoveries, including a new approach to physical therapy.

You could toss virtual balls in VR using a first-generation DataGlove, but only slowly; the balls moved in slow motion as well. This was an advantage for klutzy jugglers—make the balls slow enough and anyone can juggle. We realized that people could learn to juggle real balls by gradually speeding up the virtual balls. It's a way of removing a hump from the process of learning a physical skill. Make it slow and easy in VR and then gradually make it more speedy and realistic. This idea is now commonplace in advanced rehab. For instance, there are therapy systems that use slowed-down VR to help recipients of artificial legs get accustomed to them.

Fifteenth VR Definition: Instrumentation to make
your world change into a place where it is
easier to learn.

First VR Consumer Product

We were always trying to find a way to get cheap VR gadgets out to the masses. The best-known example was a glove. Once VPL turned into a real company—and that story is coming up—we had a deal with the giant toy

The Power Glove, for an early Nintendo game system, foreseeing Wii and
Kinect as a mass-market haptic input device that worked with a game console.

company Mattel to sell the Power Glove, which worked with an early generation of Nintendo game machines. Millions of these gloves were made. I wish more people could have seen the prototype games and experiences. They would still impress today. But what shipped was, as always, a compromise. The device is remembered more than the official games released for it.

The Power Glove nonetheless won a life of its own in pop culture. It has its fetishists to this day. Power Gloves are cute!

Sixteenth VR Definition: Entertainment products that
create illusions of another place, another body,
or another logic for how the world works.

Interspecies Gastric Haptics

Early VR gloves were not just cute, but evocative and inspirational. They were props for bizarre and sometimes funny experiments. One example unfolded on the ridge above Silicon Valley where Penny Patterson, her Gorilla Foundation, and Coco the talking gorilla resided.

Coco apparently used sign language to communicate, but gestured with extreme, blurring speed. There was a controversy about whether her signs were being interpreted correctly. Were researchers reading in more than Coco could really intend?

So, Penny asked if we could make a DataGlove for a gorilla. Of course! Great market for a struggling company. What was I thinking?

I drove up to the Gorilla Foundation with an expensive, self-funded test glove, but as it happened Coco was in heat. My presence was a little too hot for her. Instead, Penny suggested we try the glove on the younger male, Michael.

Penny pulled the glove down on Michael's hand, and he looked at it quizzically for about a second before he gobbled it whole in one thunderstruck gulp.

Months later I got a call. "Remember that glove? It's come out." Apparently the gorilla digestive tract can semipetrify objects that can't be digested. I wanted it! What an object. We argued. Alas, the Gorilla Foundation retained it, and possession being nine-tenths of the law, I can't show you a picture.

Octopus Butler Robot

What about active haptics, meaning devices that don't just sense how your body is moving but can convey force, resistance, heat, sharpness, or other sensations of touch?

Haptics experiments in the 1970s had involved rather huge and scary robotic arms that could be programmed to convey events in a virtual world. Fred Brooks worked with examples of these pieces of big iron in Chapel Hill. They'd often be attached to the ceiling, just like Ivan's early display rigs.

A robotic arm is active; it can convey when a virtual object is an obstacle. You move the arm, the arm moves a cursor or virtual tool, or perhaps even an avatar hand. When that virtual extension of your hand reaches an obstacle, like the surface of a virtual desk, the robot refuses to move through it. It feels like you've hit a surface because of actual sensations, not inferred or synesthetic ones. Your brain weaves together the haptic cue from the robot with the computer graphics imagery of the desk that you're seeing, and unless you're being ornery, you'll experience the physicality of the desk.

If the haptic gadget functions well, then when you instead attempt to punch a virtual bean bag chair, you'll feel a soft, crunchy give rather than the crisp finality of the desk surface. Similarly, if you pick up a virtual weight, the robotic arm can pull downward on your real hand, to simulate gravity.

This is called *force feedback*. I have made it sound easier than it is. There is always a tremendous challenge with reducing latency and improving accuracy, just as there is with vision in VR, but that is only the beginning of the problems. One has to figure out how to anchor the robot, and then, crucially, the contraption ought to not ever harm you, even if programmed stupidly.

The perception of force feedback is fascinating because it uses the whole body. When you press down against the top of a desk, whether real or virtual, your whole body feels it. If you are standing, your whole frame adjusts to both sense and cope with the reality of that desk resisting you. If you are seated, your whole arm and back adjust. You sense your body's pose and the strains that influence it with that portion of haptic experience called *proprioception*, as well as with the tactile sensation coming from the local area where you are pressing.

Force feedback is one of those subspecialties of VR that's had a commercial life out in the world for many years. This is a personal book, not an account of the whole field, so instead of mentioning everyone, I'll highlight my favorite force feedback researcher, Stanford's Ken Salisbury. One of the devices he coinvented, the Phantom, has been a common building block in VR systems for years. It's a cute, desktop-friendly robotic arm that a single hand can use to operate a virtual instrument.

Force feedback devices along these lines are commonly applied to medicine. You can pretend that a penlike extension is the handle of a real device, such as a scalpel. This is exactly what is done in surgical simulators.

A surgeon once let me perform part of a laser procedure on my own retina; I had helped design the instrumentation. Of course the doctor was *way* not following the rules, so I won't mention names.

As wonderful as force feedback devices are, there are manifest limitations. For one thing, they have to be anchored, so it's hard to move around and still be able to use them. So one imagines mounting force feedback devices on robots that quietly keep pace with you, ready to swing around to meet your hand when needed. Or maybe the whole floor scrolls so the robot can stay in place. We've tried it both ways. They're both hard.

Anyway, Ken and I, and a few other cohorts including Henry Fuchs, used to call this the "butler strategy."

Here's a little more detail about how it would work: Imagine you're in a virtual world and you want to slam your hand down on a virtual desktop. Now suppose there's an attentive robot scurrying nearby. (You don't see the robot, of course, because you are looking only at the virtual, computer-generated world.) The robot has an arm that is holding a tray, like a butler. As you start to slam your hand down, the robot calculates that you ought to hit the virtual countertop. The robot swoops just in time, bringing the physical tray into alignment with the virtual desktop and creating the illusion for you that the countertop was there all along.

Please, for a moment, put aside questions of safety; we're only performing a thought experiment . . .

If you were to run your fingers along the surface of the butler's tray, you would have to come to the edge pretty soon, since the tray would have to be small enough to swing around without hitting you . . . but desks can be big. So the butler robot would presumably have to move the tray to keep up with your hand so that the surface would feel bigger than it is . . . but then you wouldn't feel the surface moving against your fingers as you move them across it.

This brings us to another aspect of haptics, tactile sensation. This is sensation that comes from sensor cells embedded in the skin.

Tactile feedback is astonishing because it is actually a whole ecosystem of distinct senses. There are many different types of sensor cells in your skin. Some sense heat, others sharpness, others pliancy, and there are often variations that sense only changes in such qualities, not the qualities themselves.

Some sensor cells are sensitive to textures as you move your fingers over an object. Okay, take a deep breath: To feed these texture-stroking cells with the sensations they expect, the tray proffered by the butler robot will have to have a surface coating that scrolls—in any direction—to simulate stillness when the tray moves, so that it can simulate a surface larger than itself. I know this mechanism can be hard to visualize or understand from a written description. Even professional VR researchers can get stuck trying to remain unconfused by the inside-out gadgetry we have to build.

Okay, so now what if your fingers are supposed to feel a samovar, or worse, a chicken? A samovar has curved surfaces, so the robot would have to offer up places to touch that conform to and follow the curves. How could that be done?

Nature gives us clues. There are animals that can change shape rather remarkably, like the mimic octopus. So Ken and I looked into making a robotic mimic of the mimic octopus. This would be a robot that could morph into myriad shapes quietly and quickly.

It would take on the shape of the part of a samovar you are about to touch. Your brain would believe the whole samovar was there.

Certain elite cephalopods can raise patterns of welts in order to take on different textures for camouflage. We're starting to see experimental artificial materials that can perform a little of this elaborate trick. Hard metal is relatively easy to simulate. Could a material distort itself so as to feel like the side of a chicken? Feathers and all? Maybe someday.

So, you can see that all the components for a general scheme for active haptics are becoming at least imaginable. We used to have a long-term plan to make an "octopus butler robot" that would present such a wide range of haptic feedback that you could just let your imagination loose, as we already can with the visual side of VR.

But . . . what a pain in the ass. None of us have had the patience to really see the whole agenda through. Please know that I presented quite an oversimplified version of it.

The problem with active haptics—meaning haptics that pushes back at you—is that it gravitates away from generality. You can come up with a design like the Phantom to simulate the feeling of holding a scalpel, but it's hard even to imagine devices that could anticipate the forces and feelings of a wide range of virtual worlds that could work wherever you might like.

Generality is part of the core idea of VR.

Seventeenth VR Definition: A general-purpose
simulator, as compared to special-purpose ones
like flight or surgical simulators.

I've been talking about haptics in full-on classical virtual reality. In the mixed reality variant, in which you still see, hear, and feel the real world—and see/hear virtual stuff added to it, the situation is different. In that case the software can find physical affordances in the environment as impromptu props for haptic feedback. You can put a virtual slider along the edge of a real desk, for instance. That makes it easier to enter values on a fake slider than when moving the hand in the air. It lets you adjust the slider steadily and accurately, and avoids the problem of arm fatigue.*

But back to classical VR. Unfortunately, active haptic devices often demand that VR become specialized to uses that involve particular hand tools. The kinds of active haptic gadgets I described also tend to make VR less ambulatory, since the contraptions usually have to be anchored to the physical world, like cranes. For these and other reasons, active haptic devices often confine VR to specialized applications, where it isn't really VR anymore.

But passive haptic devices don't have that problem.

Lick the Problem

The hands are not the only output devices of the human body. VR is all about measurement.

Of course one might speak in order to change a virtual world. Speech is poor at effecting continuous change, however, though voice needn't be. Maybe singing will become paired with speech in a new way in order to combine discrete and continuous aspects of virtual world interaction.

The highest-bandwidth inbound sensory pathway to the brain comes from the eyes via the optic nerve. The highest-bandwidth outbound pathway to a single organ goes to—do you know?—the tongue! The tongue is the only part of the body other than the face that can be deformed continuously in any significant way, instead of being primarily jointed, like arms or legs. Unlike the face, the tongue is rather underutilized much of the time. If you're not eating or talking, there it sits.

I have experimented with using the tongue as an input device for years and have come to believe it offers unique potential. It isn't easy to sense the tongue's shape without putting sensors in the mouth. I have tried ultrasound scanners, similar to those used to observe fetuses. There are other ways; much of the research has been done to try to improve interfaces for people with paralyzed bodies. It's been done with tooth implants, tongue

---

* Must mention Columbia's Steve Feiner, who has done wonderful work along these lines, pardon the pun.

implants, and disgusting removable devices that are similar to removable braces but less comfortable.

People can instantly learn to control interfaces with their tongues. They can control multiple continuous parameters at once, for instance, like an octopus running a mixing board. Tongue agility varies, but most tongues can morph enough that they might someday be the best way to guide the process of geometric design in virtual worlds. It's also easy to learn to use teeth as buttons, if one really insists on having buttons.

Deepest Time Machine

In the earliest experiments with networked virtual reality, each person would appear inside the simulation as a floating head and hand only. It was all that the computers of the day could manage, since we had to reduce the amount of visual detail in a virtual world to the bare minimum if we wanted it to run fast enough to be usable.

As soon as computers got fast enough to show whole avatars, we built full-body DataSuits so that people would be able to drive avatars with their whole bodies. These were probably the first motion-capture suits sold. (This type of suit is still sold today, most often to capture an actor's performance to drive an animated character.)

It came to pass that nonrealistic whole-body avatars were occasionally created by mistake. These would usually result in a complete breakdown of usability. As an example, if an avatar's head were made to jut out of the side of the hip, the world would appear rotated awkwardly to the user, who would immediately become disoriented or worse.

In the course of exploring avatar designs, we occasionally came upon an unusual body plan that worked for people—no vomiting—despite being nonrealistic or even bizarre. The first example I recall was the one that opened this book: when my hand became huge over Seattle.

We naturally undertook an informal study of "weird avatars that were still usable." We took turns occupying a series of increasingly strange, but usable, nonhuman bodies. Most of these were at least mammal-like, in terms of overall structure and inventory of limbs.

The ultimate strange avatars departed from the mammalian plan. Ann had seen a postcard of people in lobster suits, part of a festival in a Maine lobstering community.

She created a lobster avatar. Since the lobster body includes more limbs than a person, there were not enough parameters measured by the body suit to drive the lobster avatar in a 1–1 map. We had to map the degrees of freedom of the body suit to the greater number of degrees of freedom of the lobster. We found tricks that worked. For instance, moving one's left and right elbows together conveyed bend information to the pereiopods with more intensity than would be the case when they were moved out of sync.

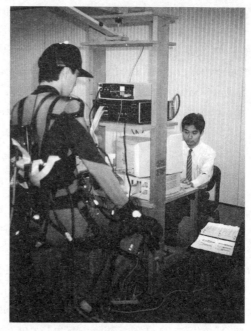

DataSuit in development.

Through strategies such as this it became possible to map human to lobster. What was most surprising was that most people could learn to be a lobster with relative ease. I found it easier to be one than to eat one.

I dubbed the study of weird avatars *homuncular flexibility*. The homunculus is the mapping of your body to your motor cortex, and is visualized as a deformed creature spread across the surface of the brain. (Yes, I know it would have been better to call it homuncular plasticity, but it was a late night.)

Homuncular flexibility is a deep topic that I can touch upon only briefly here. I'll at least mention that it's interesting to give a person a third arm, but in that case the illusion can be pushed further than you might imagine.

Certain haptic illusions can create sensations that seem to come from *outside* the body. You can put buzzers on each arm of a subject, and after tedious tweaking and proper stage setting, your subject will possibly feel a buzzing sensation coming out of the thin air from between the arms. It's a spooky feeling.

If you ask the subject to put on a VR headset and then implement a visually perceived third arm in the same location that the buzzing seems to be coming from, then the haptic perception no longer feels as if it is coming out of thin air, but instead from the third arm. Thus, active haptic feedback can, to a limited degree, be incorporated into virtual limbs without having to engage the brain directly with electrodes or energy beams. Antiphantom limbs can be implemented with conventional VR garments.

Our work with homuncular flexibility has been mirrored by the research of V. S. Ramachandran and others who study the phenomenon of phantom limbs. Rama has been able to use remarkably simple setups with mirrors to investigate cognitive phenomena similar to what we observe with elaborate VR setups.

Caltech biologist Jim Bower once commented that the range of usable nonhuman avatars might be related to the phylogenetic tree. Maybe the brain was cruising through hundreds of millions of years of deep evolution, remembering, as it were, how to control body plans of the creatures it evolved through to reach us. Maybe usable weird avatars foretell creatures the brain might be preevolved to inhabit in the deep future. We might just as easily be exploring preadaptation as prior adaptations for creatures we have the potential to become hundreds of millions of years in the future.

Eighteenth VR Definition: Instrumentation to explore
the deep time of nervous system adaptations
and preadaptations.

Haptic Intelligence

I've always thought that when VR matures someday, then an artwork, a lesson, or a conversation in VR will *not* be a made up of a virtual place you visit, as current imagination usually holds, but a form you turn into. After all, there is no absolute distinction between avatar and world in VR. If the clouds turn when you turn your wrists, you gradually take them on as part of your body map. You and clouds, one.

This would present a wide-open wilderness to explore. The most intense experiment I recall was trading eyes with another person. That is, each person's point of view tracked the other's head/eye position. The sensorimotor loop became a figure eight. It's hard at first to coordinate. The feeling can be close, sexual.

Avoid the utopian fallacy! In this case, I am not suggesting that experiences of shared or entwined avatars will necessarily lead to spiritual or even erotic altitudes. (I probably did suggest such things when I was in my twenties. In my defense, wouldn't it actually have been less forgivable if I didn't think that way? What is youth for?)

One might, while entwined in a cognitive figure eight, be looking at oneself through the other person's eyes. When we tried these things, the graphics quality was still in its origami infancy, so that wasn't significant. The haptic aspect of coordination *was* the experience.

But today, an exercise in coordination that ideally would lead to remarkable empathy and sympathy could just as well become a narcissism magnifier. One of the sharp comedians—Stephen Colbert—called the figure eight sensorimotor loop experience "fucking oneself." This is how VR will be, like all the media before it: capable of amplifying both the best and worst in people.

For better or worse, if what we're after is intensity and exploration, then let us deemphasize the idea of visiting strange places and start modifying our sensorimotor loops. When you move like a cat, you think like a cat. The brain and the body are not fully separable. When we weave new bodies in VR, we'll also be stretching our brains. This will be the heart of the VR adventure. The most profound meaning of usable weird avatars might turn out to be in the awakening of the vast part of the brain that is connected to the body.

People think differently when they express themselves physically. Like anyone else who has learned to improvise at the piano, I've been amazed to find that my hands solve mathematically rich problems in harmony faster than I could think about them in other ways.

Having a piano in front of me makes me smarter by applying the biggest part of my cortex, the part associated with haptics. The motor cortex doesn't usually deal with abstract problem solving—just concrete tasks like balancing and catching balls—but piano improvisation proves that it's possible.

I've always been fascinated with this potential and have tried turning kids into avatars such as DNA molecules and abstract geometry problems inside fancy VR systems that allow for full-body interaction.

Don't just think of VR as the place where you can look at a molecule in 3-D, or perhaps handle one, like all those psychiatrists in Freud avatars. No! VR is the place where you become a molecule. Where you learn to think like a molecule. Your brain is waiting for the chance.

Nineteenth VR Definition: Instrumentation to explore motor cortex intelligence.

As If One Obsession Weren't Enough

All that talk about pianos and haptic intelligence brings us to a personal eccentricity that blossomed in me once I had my own home in Silicon Valley. I call it "organomania": a need to always be learning to play a new musical instrument.

I had brought two objects from my parents with me to New York and then to California. One was Ellery's Royal portable typewriter. The other was the painted Viennese zither from Lilly.

In New York I had found a cheap plastic shakuhachi, cast from the classical Japanese bamboo flute, and learned a little about how to play it from Teiji Ito, who had been married to my favorite filmmaker, Maya Deren. On arrival in Palo Alto I therefore had three instruments: shakuhachi, clarinet, and zither. I also rented a little upright piano.

But then a slow-moving disaster unfolded. The shakuhachi was incredibly thrilling to play, and I couldn't stop there. I had long been enchanted by what was then called "world music." When I was a kid we had 78s of Uday Shankar and other amazing non-Western musicians. In the LP era I was obsessed with releases on the Nonesuch label: gamelan, Tibetan ritual, drumming from Ghana and Senegal, gagaku court music, and on and on.

Each time I heard a new variety of music it felt like a hidden cave within me opening up.

The Bay Area, it turned out, was at the time one of the most international cultural spots. There were intense Chinese music clubs in the basements of San Francisco's Chinatown. Ali Akbar Khan started a premier school for North Indian classical raga in Marin. There were gamelans, West African drum troupes, Taiko dojos, flamenco cafés.

So I studied everything I could. And that meant that instruments started accumulating in my hut. A lot of instruments.

Around 1982 there might have only been a few dozen or so. That's what it looks like in old photos. A girlfriend would say, "Can you at least keep instruments off the table? I'm afraid to move them and I want to eat."

Organomania is apparently incurable. Today there are well over a thousand instruments in our home, maybe two thousand, and I have learned to play each one at least well enough to enjoy. Not as impressive as it might sound, since a lot of instruments are similar to one another, but still, my obsession has consumed much of my life.

I always say, "At least it's cheaper than heroin," though I'm not really sure if that's true. The story of the instruments will have to wait for another book, but they appear in this narrative because they are central to how I appreciate VR.

There are uncelebrated cultures of haptics all around us if we take a moment to feel, but my favorite is found in musical instruments.

When you learn to play an instrument from a faraway time and place, you by necessity learn to move your body in ways that are at least related to the ways of the original players. Instruments provide a haptic channel across centuries and continents, like writing, but less symbolic and far more intimate.

Certain musical instruments convey strength and force. Various horns, bagpipes, and drums were battle tools, practically weapons. You play these with the biggest muscles and you must gird yourself in order to focus power. The size of the muscle groups you use is deeply connected to the rhythms you'll tend to play. Lost music comes alive, just a little, between your body and an ancient instrument.

Other musical instruments are measured to the human body, rather than an external, worldly situation, so that you can play them with minimal motion, approaching a trance. The oud is like this. There are wind instruments that make you notice ever more subtle aspects of how you play, even after years, like the shakuhachi, and others that make you want to be ever more accurate, speedy, and showy, like a modern flute. You feel the difference in your throat.

Not only are instruments the best haptic interfaces yet invented, they are the best interfaces of any kind, if what we care about is the potential for mastery and expressiveness.

Instruments show what's possible; how far computer science must go before we can call it even a beginning.

Harm and Heal

Homuncular flexibility has been studied in labs around the world for decades because it's a great way to understand the relationship of the brain to the rest of the body.

I must mention two special researchers: Mel Slater (University of Barcelona and University College London) has conducted wonderful experiments, including testing how well people can learn to use tails if tails are grown on their avatars.* The answer is that people are great with tails. We lost them to evolution only recently, and our brains aren't surprised at all to find them back in place. This is only one random example from a long career; I can't do Mel's work justice here, but look him up.

Jeremy Bailenson of Stanford is particularly dear to me. I've worked with him since he was a student, and he's now running an ever more amazing lab that studies everything about avatars.† His work is brave and haunting. He's studied how people perceive each other differently when avatars change. Alas, social status rises as a person's avatar becomes taller, and yes, his work has much to teach us about racism and other sad facets of our character.

Jeremy and I have embarked on a long-term quest to map out the range of possible avatars. What creatures is the human brain adapted or preadapted to inhabit?

Sometimes a student of Jeremy's becomes an intern in my lab. I have to mention one recent and exciting case. Andrea Stevenson Won, then a student, now faculty at Cornell, prototyped a pain management app using avatars in 2015.

Here's the idea: Patients in chronic pain draw virtual tattoos where it hurts, and then interact with other people in mixed reality so that the

---

* http://publicationslist.org/data/melslater/ref-238/steptoe.pdf.
† http://onlinelibrary.wiley.com/doi/10.1111/jcc4.12107/full.

tattoos are socially acknowledged. Then a therapist causes the virtual tattoos to gradually spread out and dissipate. Her work might be a path to reducing the subjective intensity of chronic pain.

It's also a way of applying the evil effects Jeremy has documented to good works. Yes, we can use VR to make people more racist, fearful, or subservient by tweaking avatar design, but maybe we can also potentially manage pain better.

I can't emphasize enough how young the science of VR remains. We still know so little.

12. Nautical Dawn

Other Sighs

During the period of half-light, when we were building experimental phenotropic programming languages controlled by gloves, but before computers were good enough to prototype social VR in proper 3-D with goggles, there was nothing official binding our loose tribe together.

Tom, Ann, Young, Chuck, Steve, and a shifting cast of other curious souls were fiercely interested, but not exclusively. We'd all drift in and out. Ann and Young still were finishing their degrees at Stanford, while Chuck, Steve, and Tom had various freelance projects that paid the bills.

So did I. Before the first VR company, VPL Research, was incorporated, there was another little effort, my trial run at having a startup. It was a collaboration with Walter Greenleaf.

Walter was working on a neuroscience PhD at Stanford, and it so happened he had the best pickup line on campus. "Would you like to be a test subject in the Stanford sex lab?" Subjects were directed to achieve orgasm while wired up with sensors.

This lab was reached by going up a spiral staircase under a faux Renaissance dome in a decrepit but ornate old building that had long ago been used for formal receptions. Then you went through bead curtains, past psychedelic posters. The med school stored cadavers in the basement. The building became notorious as a one-stop shop for the study of the sacred and mystical seams of human existence. Sex, sleep, and death.

Sadly, the site was soon to be bulldozed and forgotten. A parking structure now occupies most of the ghost of its footprint. (It's the multistory one east of the Gates computer science building, if you're curious.) I always try to park as close as I can to the place where the dome used to stand.

This was the same lab where, around the same time, Stephen LaBerge demonstrated that lucid dreaming was real.

Lucid dreaming means you become aware within your dream that you're dreaming. With practice, you can direct the events of the dream. You can fly or will diamond palaces into being. It can feel not only "real," but realer than real, even though you know it isn't.

People would fly, of course, and have superhuman sex; they'd cause sea monsters to arise at the scale of mountains. How could Stephen test whether lucid dreams were really happening, that the effect wasn't just people telling stories?

In one of Stephen's experiments, subjects were asked to move their eyes around in a predetermined pattern from within a lucid dream. (The eyes still move during sleep, even though the rest of the body is mostly immobilized and disconnected from events in the dream.)

When he was able to observe people in confirmed REM sleep moving their eyes as planned, he had shown that they were in control from within the dream state. Stephen could also use various scandalous sensors to measure whether subjects really experienced having sex when they said they did.

One of my hacker buddies dated a woman who had achieved more orgasms from within lucid dreams than any other subject in the lab. (He pointedly did not ask her what was going on in the dreams while this happened.) Quantitative boasts are silly, but isn't that more interesting than counting fake friends on social media today?

I learned to lucid dream and found it briefly fascinating and then boring. All by yourself, you make up experiences only for yourself. Being able to remember dreams, to let the brain run free, is more interesting. But beyond that, the point isn't the contents of reality, but the connection to other people. Scale isn't real. A giant crystal dragon might as well be a word, unless people can do more with it than with words; and that can only happen in collaboration.

Walter, Stephen, and the informal crew around my hut collaborated in various ways. We put VR gloves on lucid dreamers to see if any faint movement from within the dream might be detectable—answer at that time: not reliably enough for science.

Other Everythings

In the early 1980s, a lot of people thought of lucid dreaming and VR as twin research projects. I'd often be asked to compare them.

Twentieth VR Definition: Like lucid dreaming, except that (a) more than one person can take on roles in the same experience, (b) the quality is not as good, and (c) you have to work to program VR if you want to be in control, which you should want. Dreams, meanwhile, are often best if you don't seek to control them. Even Stephen LaBerge seeks to be nonlucid in most of his dreams, since it is in untethered dreams that the brain surprises and renews itself.

Lucid dreaming was actually only one of three parallel "dreams of everything" burbling forth at once. VR was another, and nanotechnology was the third.

At the time, Eric Drexler was popularizing the term "nanotechnology" as a psychedelic program to reformulate physical reality into the varieties of unbound scenarios we might hope to realize within lucid dreams or VR. Instead of accepting that physical reality was forever to be mostly made of objects not of man's invention, like stars and rocks, we would learn to control reality by commanding atoms into any configuration we might fancy. We'd fly into space and our skin would produce a golden film to protect us from the vacuum; we'd will ourselves into being great space beasts and paradisiacal bubble gardens.

Lately, "nanotechnology" has a sobered-up definition, meaning a certain spirit of ambitious chemistry, such as the creation of tiny motors.

But back then I was often asked to compare it to VR. I was a little less charitable making that comparison.

Twenty-first VR Definition: In comparison to older, grandiose definitions of "nanotechnology," VR lets you experience wild things without messing up the one physical world that others are compelled to share with you. VR is vastly more ethical. It's also not so nutty. We can see how VR will work without weird speculations or apparent violations of fundamental physical laws.

I got a lot of pushback! People would say, "Why would you go to all the trouble of programming virtual worlds when—any day now—you will be able to change physical reality to actually conform to any possible virtual world?" Same old presumption of supremacy.

But I don't want to dismiss the line of thinking completely. Maybe VR can give us a hint about what we should want in the future as technology ramps up and people gain more and more options. We can simulate living in a Jetsons world today. Let's try and see if we really want it.

Twenty-second VR Definition: A preview of what reality might be like when technology gets better someday.

The Previous Everything

The eight-hundred-pound flying gorilla in the room was psychedelic drugs. They comprised the extant "dream of everything" that captivated baby boomers. I am usually binned with the next cohort, known as Generation X, but the boomers were dominant and laid down the context for everything we did.

The question I was asked more than any other in the early 1980s was how VR and LSD were related. Once it became possible for people to actually try VR, the question finally started to fade. But for the record:

Twenty-third VR Definition: VR is sometimes
compared to LSD, but VR users can share a world
objectively, even if it's fantastical, while LSD users
cannot. VR worlds will require design and engineering
effort, and will be best when you are willing to make
the effort to create and share your own experiences. It will be like riding a bike, not a roller-coaster
ride. Although there will be thrilling VR experiences,
you'll always able to take off the goggles. You won't
lose control. VR will typically be "lower quality" than
reality or dreams or psychedelic trips, although it
will be up to you to hone your senses so you can
notice the difference. LSD is ready now and VR won't
be really good for a while. It might be more for your
kids or their kids.

Timothy Leary found a new cause célèbre in the idea of VR, though it wasn't yet possible to try it. If you're too young to remember, Tim was earlier known as "the most dangerous man in America" for his infectious revelries, which involved not only psychedelic drugs but also proclamations that everything was suddenly different; that it was best to reject and ignore old institutions like government, college, and money.

He saw the world as being on the cusp of a revelation, after which we would all experience more peace and beauty, and he thought drugs were a key to making it over that hump. He was one of the most influential figures in baby boomer culture, and he helped define a cultural divide that remains torturous in America today.

On one occasion, Tim declared VR to be the new LSD. We disagreed strongly on this matter.

After a round of back and forths in underground 'zines and the like, Tim asked me to meet him privately so we could hash out my concerns. Of course the meetup plan was devious and delicious.

Tim informed me that I was to sneak him out of the Esalen Institute in Big Sur, where he was contracted to give a workshop. First I was to pick up and smuggle in a professional Tim impersonator who would take over the workshop duties. Then I would conceal Tim in the trunk of my car and casually sail out past the guards at the gate. We were to enact a noir movie set in Cold War Berlin. Sure, why not?

Since it was me, clearing out the trunk of my car was a challenge. Walter helped me excavate it next to a dumpster behind his Stanford lab. We triaged towers of printouts, computer tapes, and floppy disks the size of placemats. We dumped a few computers I hadn't booted in a while that would make nice relics today, just to make room for Tim. There was an Apple III, a prototype from Sun, and part of a LISP machine.*

My heart thumped as I tried to not look guilty, avoiding eye contact, waving meekly at the guard in the booth. I glanced momentarily and saw that instead of the muscular, uniformed enforcer I feared, there was a tiny, spacey, bearded young man in a tie-dyed T-shirt.

The look-alike was a success, and so far as I know, no one detected the deception. Presumably the students were dosed. The extraction scheme worked!

When the time came, Tim did fit in the space we carved out, though some remaining equipment immediately fell on him. Would the mission succeed? Tim had given me the address of the most spectacular house on Big Sur's Pfeiffer Point, and once there I got him out of the trunk and we dined with Hollywood people high above the waves under the full moon, listening to rather stunning unreleased Talking Heads tracks.

Tim was always ringed by circles of adoring hippie kids, but he also adored being embedded in the world of Hollywood glamor. He became a great friend with whom I disagreed. It was good practice for me, as I would grow to have more of those over time.

Once I found myself speaking at a conference in Spain where Albert Hofmann, LSD's inventor, also spoke. He came up to me and said, "You have inherited Tim." A sly glance. I was speechless.

Tim and I never reached an agreement about how to compare psychedelic drugs with VR. He did agree to tone down the rhetoric about VR just a little, which was helpful. The last thing we needed was a massive allergic response against VR before it was even working.

Through Tim I met the rest of the psychedelic world. I was particu-

---

* LISP machines were computers dedicated to, you guessed it, LISP, an early computer programming language beloved by mathematicians and artificial intelligence researchers of a certain age.

larly fond of Sasha Shulgin, the astonishing chemist who invented and tried hundreds of new psychedelics under a special license from the U.S. government in his world-class chemistry lab hidden in a little rustic cabin in the hills behind Berkeley. He was one of the clearest-headed and sweetest people I ever met.

A subset of Tim's followers would wax on and on about how one drug would foster empathy, another joy, and the lot of them a guarantee for world peace, spiritual fulfillment, and everlasting genius. They often thought of drugs as beings, in the way that computer scientists might choose to think of computers as being alive, artificial intelligences. A psychedelic molecule in a mushroom can be cast as a creature bringing wisdom to mankind. (I must observe that psychedelic researchers got into poisonous, petty disputes about attribution, grants, and the rest of the usual loot of the science life, so the utopian force of the drugs couldn't have been all *that* powerful.)

Psychedelic utopias have an automatic quality that later turned out to merge well with the technolibertarian sensibility. Gone was the old Marxist (or Ayn Randian) sense that one needed to struggle toward utopia.

There were ways that psychedelic thinking lent maturity to my VR idealism. Behind the utopian curtain, more interesting ideas could be found, like "set and setting," meaning that drug molecules didn't *really* impose any particular meaning, absent context. For instance, MDMA (Ecstasy) was understood to be simply pleasurable, or an empathogen (stimulating empathy); it later found its broadest role as a stimulant and sensory enhancer in sleepless Euro-thumpy dance clubs. Now it's being tested in the treatment of PTSD and even autism.*

So one psychoactive molecule can have a wide range of meanings. While I never thought VR was anything like a drug, the "set and setting" principle applies to VR at least as well. VR can either be beautiful art and sympathy or terrible spying and manipulation. We set its meaning.

LSD was common in tech circles. Steve Jobs would go on and on about it.

I used to come under tremendous social pressure to use drugs, LSD especially, and pot at the very least. As it happens I have never tried them,

---

* It also might be dangerous; I am not advocating its use. Indeed, a parent of one of the Veeple died from a cardiac event while using it.

not even marijuana. It was burdensome to have to constantly explain myself. My choice was taken as an affront.

My intuition was that drugs weren't for me. Simple. Am not judging anyone else. The recent pressures one feels to join social networks feel similar. My answer is the same.*

Some people called me a liar. Supposedly I evidenced having "seen things and known things" that could only be accessed through LSD. I was a pretty freaky and psychedelic guy, I suppose. Tim Leary had a nickname for me: "the control group." I was the only person around the scene who had not taken drugs, so maybe I was the baseline. Maybe drugs made people straighter.

Someone needed to be the control group. Many years later, when Richard Feynman knew his cancer was starting to overtake him, he decided it was time to experiment with LSD. The plan was to hang out with some hippie women in a hot tub at the edge of an unfenced cliff high above the waves in Big Sur.† He asked the control group to be there, keeping a discreet distance to make sure he didn't fall to the rocks. The man was hilarious on LSD. Couldn't do arithmetic anymore. "The machine's broken," he said, pointing to his head in delight.

There was one drug that resonated in particular with VR: an Amazonian compound called ayahuasca or yage. William Burroughs wrote about it, and there were other famous accounts.‡

The culture around this drug finds that it creates a psychic link

---

* Instead of pointing fingers at other people (the present leader of the free world, say) who might not be at their best when they use social media, I'll reveal that I decided that I can't handle the stuff as it exists today. I don't have any social media accounts, even though I have books to promote and other motivations. I have experienced myself becoming petty in online exchanges. I've gotten into feedback loops with people who either love my work or hate it, and we've both been driven to extremes when that's not what I wanted. I fear social media would bring out the worst in me.

Maybe social media would help my career, but that's not a good enough reason to do harm to my character. I'm not saying it's necessarily bad for everyone. Maybe it's like alcohol, fine for some people, but some of us should avoid the stuff.

I worry about the amplification of alt-right paranoia, but this isn't a partisan observation. Criticism of the left from the right has often centered on how cranky college students have become. So sensitive! So quick to take offense! Recognize a pattern? The personality dysfunction that's derided by critics from the right as "poor little snowflake-ism" is the same as the one seen in President Trump. There are social media addicts all over the political spectrum.

† Not Esalen. Sorry, will not disclose the location.

‡ *The Wizard of the Upper Amazon* was the favorite in my circles, though the extremely influential books of Carlos Castaneda—ostensibly about a different drug in a different region—might really have set the tone.

between people, in which users share experiences as a form of communication that transcends words. Therefore, ayahuasca was understood in a way that was similar to the way I thought about the future of virtual reality.

The similarities went beyond that: Both could make people throw up. That's not just a flippant comment. Both demanded an element of risk, preparation, and potential sacrifice. Perfect setup for ritualistic adoration.

VR rarely makes people throw up these days—we don't even have barf bags around demos anymore—but fascination with ayahuasca culture, which has recently enjoyed an improved legal climate in Brazil, continues to draw VR engineers. A batch of Silicon Valley VR-heads go down regularly, and there are still events in California that attempt to re-create the Amazonian rituals.

I never tried ayahuasca, so I will reserve judgment about what it does. I will say that I have never seen evidence of psychic links between people who use it, and I have been around people using it more than a few times. So I walk the tightrope I often talk about. You know, the one where if you fall off to the left you're superstitious, and if you fall off to the right you're reductionist.

Stim City

Anyway, back to Palo Alto, around 1982.

Walter and I used sensors related to the ones in Stephen LaBerge's rig to create a simple monitoring device for vital signs. A partial glove. Put it on and you saw a real-time representation of your guts on a screen. Lungs matching yours; you could check by breathing in deeply and watching them expand. Same for your beating heart.

Data was recorded, though that was mostly mocked up, because it was way too expensive to actually store much data in those days; plus a bored employee would have to sit there constantly exchanging floppy disks.

Our idea was that we'd gather a bunch of data from people and eventually algorithms might find correlations related to health. Maybe the system would become able to diagnose illness. Maybe it would help people learn to control their stress or track their fitness. A toy to make you healthy!

This should all sound familiar, since devices like fitness bands are everywhere—and often oversold—but at the time the idea was fresh and startling.

Walter and I used to collaborate nocturnally, which had something to do with the schedule Walter kept at the sleep lab. I'd bundle up a computer, usually an Apple Lisa with luggage straps I'd added, and we'd work in an all-night diner. There were just a few with wall outlets in usable places; you had to be tactical to get the table you needed. "I'll pay for your ham and eggs if you trade seats!"

One night we were working at what I'll call the "creamery" near Stanford. Just to be clear, this is NOT the same joint called a creamery that is found in Palo Alto today. The clarification is important because the overnight entertainment was watching the owner shriek like a ninja as he attempted to spear the rats that ran behind the counter. We never saw him succeed, but admired his determination. A few of the more spirited rats had names and were spoken of affectionately in hacking circles.

"It's amazing how persistent this guy is, even though he never gets a rat."

"If he was working in tech, he'd have a mega company by now."

"Why don't we try?"

We bundled up a prototype and drove to the Consumer Electronics Show in Las Vegas to demonstrate our device. Maybe a big company would want to license it from us!

We were naïve. We got hooked up with a not entirely reputable business partner who was supposed to make connections and close deals. Actually, he put us up in a fleabag porn hotel and didn't accomplish anything.

But we did learn a bit about how the world works. Walter remembers an enthusiastic reception. I remember potential customers grossed out by seeing animated representations of their guts.

I also remember feeling the joy of entrepreneurship. Invent. Bring it to people. Enjoy. Repeat.

In the 1990s, after VPL's demise, Walter became interested in VR as a tool for research and treatment, especially in behavioral medicine. He's since used VR to work with gang members on violent impulse control, and other fascinating applications. And in the new century he introduced me to my wife. "She's like Betty Boop" were the first words I heard about her, and they were true.

Legitimacy, Hair, a Giant's Shoulder

It sounds ridiculous now, but around age twenty-two I was enjoying all I have described, and yet I also feared that I was an irredeemable failure. I was ashamed that I had blown my education. My mother would have wanted me to become a Harvard professor, I imagined. An anachronistic notion took hold of me that I had to find a path to legitimacy. I wanted to be invited into one of the castles that Silicon Valley might burn to the ground.

The company that sold Moondust asked if I could present it at the preeminent computer graphics convention, SIGGRAPH. This conference straddled industry and academia, so I wondered if being there in an official capacity might give me an opening.

That year's SIGGRAPH, in Boston, turned out to be nutso and exuberant. It was one of those countercultural gatherings that was still small enough to get away with genuine chaos, like the first few years of Burning Man. Also, as was true back in the hut, computers weren't fast enough to do much yet, so people had to get weird to pass the time until Moore's Law came through.

All the strands of fate clustered during my first visit to the Boston area. Before SIGGRAPH was over, I had decided to move there for a while, found a few new lifelong friends, met a woman I would eventually marry (if only bizarrely and briefly), met my most cherished mentor, and got my first real research gig.

Almost immediately I fell in with a batch of weirdo students from MIT, and it was as if we had been friends for years. Turned out they were students of Marvin Minsky, one of the founders of the field of artificial intelligence.

One of these remains a friend after all these decades. David Levitt had hair just like mine, but darker. A medium-length fountain of dreadlocks. We looked like mirror images if you squinted, though he's black, or actually "nebrew" as he called himself. He used to call me "brother from another mother."

We made for a dramatic pairing, and we cavorted ornately. Our favorite attire was vivid West African robes. Like me, David had developed a peculiar piano style, launching out of Monk and ragtime, as compared to my Scriabin, Nancarrow, and stride.

David's PhD project at MIT was in visual programming languages! Eventually he'd join the gang in California.

His parents had tilted at the radical edges of the civil rights movement. Adding rhythm to history, David has recently run for the U.S. Senate as a candidate to the left of even the Bay Area field.

This is as good a spot as any to address a nontopic that often comes up: the hair. My hair expresses no agenda other than accommodation to genetics. It is not an attempt to pass as black, or a tribute to Jamaican or Indian holy images. I simply have hyperfrizzy hair.

The ceaseless effort it took to brush it out infringed on the rest of my life, so I gave up and let it dread. That is the simple story. There's a book titled *Programmers at Work* from the early 1980s that includes a de-dreaded image of me on the cover. That's the only documentation of the brief period when I was willing to waste hours just to not look weird.

By now the dreads have grown so long that they are presenting an inconvenience in a different way and I might have to cut them. But I'm putting the question off. I don't like to worry about hair.*

---

* As long as we're on hair, I guess I can say something about my weight. It wasn't easy, but I lost the weight I'd gained as a child after about a year out of the hospital, but then it came roaring back in my teens. In my twenties I battled mightily to lose weight and did, over and over. Each time, it came back with seemingly supernatural force, and overall I found I was gaining weight in the longer term. I suspect that if I hadn't tried to change back then I would weigh less today.

Occasionally a stranger will happily tell me that I ought to work harder at it, that it was easy in their experience, and then in practically the same breath whine about how they can't get their startup funded, or their book published, or are subject to some other misfortune that seems to them to not be a fault of their own.

Silicon Valley is drowning in quantified self-help and productivity cults that are supposed to sculpt one's life into an ideal in every way. This is not only foolish but destructive. The impulse to pretend that we already understand everything is antiscience, just as surely as the antivax or anti-evolution movement. It also is a stealth conveyor of conformity. Everyone is expected to embrace the same definitions of productivity and success. And personal appearance.

There are loads of tangled, often contradictory results coming in from legitimate science about weight, as well as a cosmic, stunning amount of manipulative pseudoscience. But in reality, it's one of the many things about the universe that are not yet well understood.

Weight will probably be understood someday, however, and possibly soon, since there are so many wonderful tools for investigating biology today. When people can someday choose, they should be able to make varied choices. Diversity is an intrinsic good.

Has my weight had a negative effect on my life? In some ways, perhaps. Cameras like thin people. Maybe, if I were thinner, I'd be on TV more when the time comes to promote a book or debate the cyber-topic of the moment. But my life is about as successful as it could be, given my preferences. On some levels I might even play to type inadvertently, since smart technical people are supposed to look a little weird; Einstein with his hair. Overall it hasn't mattered much, for the simple reason that I'm male. It pains me to say that a fat woman would probably not have been able to have my career.

In those days, it was super rare for white people to have dreadlocks, so I was quite exotic. Today it's a cliché, and often not a complimentary one. Can't help that.

No one in Silicon Valley or MIT cared about my hair. But MIT was easier for me than Silicon Valley. It was like Caltech, but this time I had that stupid thing I needed: legitimacy.

Alan Kay had left Xerox PARC to start a new lab supported by Atari. He offered me a research position for the summer, a post that would normally go to a graduate student. I was in again! I had overcome my fall from grace.

Atari's lab was practically embedded in MIT, on Kendall Square. It was one of the progenitors of MIT's influential Media Lab, which would come into existence a few years later.

This is how I met Marvin Minsky, who became perhaps the sweetest and most generous of my mentors.

I've described a few things that happened while I lived in Cambridge in my earlier books, like becoming lost in Marvin's copiously disordered home and arguing with Richard Stallman about the dawn of free software. I won't repeat those stories here, but I would like you to read what I wrote about Marvin on the day he died in 2016 (this was for the tribute on John Brockman's edge.org):

The last time I saw Marvin, just a few months ago, he was hanging out in his wonderful house, front door unlocked, students dropping by unannounced. One young MIT student had worked for a summer in a circus, and naturally a trapeze hung from the vaulted ceiling. She slid upward like a cat and swung about as we all argued about AI, just like it was forty years ago.

I remembered when that trapeze was being installed, and I was the young protégé. Why was it hung there? I don't remember, but it was also when the tuba arrived in its place under a piano, now obscured by books, telescope parts, many wonderful things.

On my way to see Marvin that night I got a call from a mutual friend. "Don't argue with him, he's frail." I couldn't believe what I was hearing. "But Marvin thrives on arguments."

I was right. Marvin said, "What you're doing, criticizing AI, it's perfect. If you're wrong in the big picture, you'll make AI better. There's a lot of terrible work, after all. If you're right in the big picture, then you're right. So great!"

Marvin invented about half of the way we think about ourselves these days. His particular way of characterizing AI consumed a million imaginations. Marvin's narrative about the future of machines is the thing people are afraid of. But that's a sideshow.

The main event is that Marvin's way of thinking about people and our emotions has more or less replaced Freud's mythology. Pixar's *Inside Out*, for instance, feels and even looks like Marvin's lectures from decades ago. (He used to ask that we imagine our brains painting colors on memories of things or events so that we might react to them with a given emotion, for instance.)

And all this might be taken as an aside from his work at the foundations of computer science. And his technical contributions to so many other fields. The latest virtual reality optics have been influenced by one of Marvin's inventions, for instance: the confocal microscope.

Why was Marvin so generous to me? I gave him grief. I disagreed with him at every turn. I wasn't ever his student, officially, and yet he mentored me, inspired me, put serious time into helping me. His kindness was total, a singularity of kindness.

He came out to visit in California in the 1980s, when I was in my twenties and virtual reality was getting tolerable. He sat in a headset—was it a simulation of being inside a hippocampus with neurons firing?—while at the same time playing a real physical grand piano, and the two planes of reality became beautifully coordinated.

The music! Everyone knows Marvin improvised at the piano in the approximate style of Bach—elaborate counterpoint—but he never fell into a rut. He was just as fascinated by the obscure musical instruments I brought by, from around the world. Since everything was always new to Marvin, even Bach's style was always brand-new. Marvin lacked the capacity to become jaded or bored, or to fall into any state of mind shy of being startled by the constant novelty of reality.

I remember Marvin talking to Margaret, his daughter, and me about his take on Alan Watts. It's hard to imagine a philosopher who might seem more distant from Marvin than the gurulike, Asian-leaning Watts, and yet Marvin thought Watts was remarkably wise about death. I recall Marvin discussing Watts's idea that reincarnation is the wave way of interpreting people instead of the particle way. (Not that Marvin, or Watts, for that matter, accepted the notion of individual survival through incarnations. Instead, a person's properties or patterns would eventually reappear, approximately, in new combinations in fresh sets of people.)

I remember once we were walking near cheerful shops on a spring day in Cambridge and we came upon an infant in a stroller. Marvin started talking about "it" as if the baby were a device, a gadget, and I completely knew he was doing so to get a rise out of me. "It's able to track objects in the visual field, but with limited interaction capabilities; it has not yet built up a corpus of observed behavioral properties to correlate with visual stimuli."

Oh, that sly smile. He guessed I was the one who'd get all huffy and thus prove that I was the slave to my ideas. Marvin's warmth shone through so radiantly that the ruse didn't work. We laughed.

Marvin linked humor with wisdom. Humor was his brain's way of noticing a hole to fill, a way to be wiser. I always think of him finding a way to make each moment a little funnier, a little wiser, a little warmer, a little kinder. He never failed at that, so far as I ever saw.

Ah, Marvin.

At Atari Research there were real resources. We could print on laser printers, send email to each other, and do other digital things that were quite futuristic, elite, and exclusive at the time. I had rappelled over the chasm and was back in the world of big science.

I worked on some profoundly nonmainstream programming language ideas, as well as a couple of strange haptic games, including a robotic broomstick one could ride in a simulator. In order to be a witch. Once again, vaguely sexual.

Speaking of which, I have covered a lot of what happened in Cambridge: the new friends, the mentor, the research gig. What about the woman?

She'll remain unnamed, but her name wasn't remarkable anyway. What astounded was her presence. An iridescent goddess, a perfect archetypal blond cliché with a psychedelic hippie twist.

Flirty, weirdly wise, verbiage, cleavage, everything. Calculated indifference. I had fallen for women before, but this was electric free fall, an entirely different experience.

But here's the strange thing. I didn't actually feel attracted to her in a direct sense. It was more the case that everyone else was attracted to her, so that I lost myself in a social tide.

She was a status symbol. It felt like stumbling into an ancient magic cult, a secret society of the powerful and the beautiful.

When I first met her, I wasn't well known. I was just one of the curious, brainy, hirsute boys to be found around MIT. She was a sexual Polaris, heads always turning to follow her, the way the heads of kittens turn to track a dangling toy.

She was deeply committed to imponderable social ambitions. In our first conversation, she said, "Oh, Tim Leary sent me to Harvard so I could seduce MIT computer geniuses into the psychedelic revolution." A secret mission of historic importance!

Nothing happened between us at the time, but I would eventually end up marrying her, if only briefly. We'll come to that in due time.

Dixie-Futurism

My enchanted interlude as a legit researcher was coming to an end.

Marvin's daughter, Margaret, was doing a PhD on haptics at MIT, and she asked me to come down with her to visit the VR lab at the University of North Carolina at Chapel Hill.

The feeling of the South retriggered my subservience to mood and I found it hard to function. Slow, steamy, kudzu-coated. Polite, segregated. It felt wrong to enjoy it. A piquant preparation with an improper ingredient; vinegar BBQ of an endangered species.

Whatever one made of the region, the lab was superb. I ought not play favorites, but UNC Chapel Hill had and still has my favorite academic VR lab.

Here was Fred Brooks, an authentic Southern gentleman. Fred had run the team that created the first commercial operating system, for IBM, and defined ASCII, the way letters are represented by bits. He was one of the people who launched the Digital Age. He also wrote one of the few great books about computers, *The Mythical Man-Month*, which sensitively explained for the first time what it was like for humans to program computers.

On top of all this, Fred was a pioneering VR researcher. At the time of my first visit, Fred was particularly interested in haptics, which was also becoming Margaret's passion, so we spent a lot of time with robot arms feeling the boundaries of virtual objects.

The other pillar of the lab at UNC is Henry Fuchs, who is also one of my favorite collaborators. He's a darting genius, hardly able to talk fast enough to get his flow of amazing ideas out. His students have comprised the absolute top tier of the field decade after decade. Modern VR would not exist without Henry.

UNC probably had the fastest graphics computers of the time, and they enchanted me. Henry's group built their own computers for the visual side of VR, at extraordinary expense. Pixel Planes was one of the first computers optimized for the kinds of graphics VR would need, though on my first visit it still took a few seconds to render each frame. But improvement would surely come, according to the Law, so we all imagined living in the future.

Soon it would be time to head back to California. I didn't have any formal arrangement to return to, but it felt important to go. Something was brewing.

Ants on a Mission

"Oh wow, didn't realize you were back!"

"Oh hi! Missed you. Just now got in. It's incredible how much light there is here. And the air. You can breathe. The air around MIT in the summer is a hot dirty syrup."

Ann had what was then known as the "Seattle look." Long straight dark hair, big doe eyes. "Enjoy it while you can. This is all going away. The other dirt road in Palo Alto just got paved."

"Oh no! That's awful. But listen to them! Wow, I missed that sound." There was a cat rescue place down the road and you could hear a hundred cats meowing, a turbulent string section.

"Yeah, you're kidding yourself. It will be driving you insane in twenty minutes and I'll have to listen to you complain. Oh, and you should know you have ants."

"Oh, who cares."

"No, I mean tons of ants, rivers of ants."

Our little group had colonized the decrepit stretch of disabled orchard around my hut. Ann and Young and their kids had moved into a mirror-image hut across from mine on the same dirt road. Other people in the group lived in nearby huts from time to time.

I wish Silicon Valley still had corners of weirdness of the same caliber, but I fear those days are long gone. Next door was a woman who was one of the few top-tier female computer business consultants of the day, but it turned out she suffered from multiple personality disorder and you never knew who she'd be. There might be an acid-tongued punk rocker harassing me, or a silver-tongued power broker enlisting me to help out with a thrust into one of the big tech companies.

It turned out the refrigerator had failed and was not just colonized by ants, but had been completely filled by them, as if Archimedes had been performing an experiment with ants instead of water. I had to walk the rusty Space Age relic out back by the creek to empty it on its side. It looked like a flattened rocket that might have flown on the cover of one of Hugo Gernsback's 1950s pulp science fiction mags, the ones in which Ellery wrote his popular science pieces. The spaceship disgorged a massive invading force that had not survived the trip. Chrome caught the sun and made me squint.

Another voice behind me. The multiple personality neighbor, today sounding completely normal, which was a little disorienting: "You look festive! What's the occasion?"

Some of my Cambridge friends and I had the not-really-worth-it idea that we'd cut new holes in big colorful windbreakers from Bali so that they could be worn sideways. Two limp sleeves hung off to the left.

"I guess it's clear-out-the-ants day. No really, this is how I always look. Everyone around here's gotten more straight while I was in Cambridge; what's going on?"

"I guess that's right. I hadn't noticed until you pointed it out."

While I was gone, the hacker dress code shifted from exotic hippie to what would today be called "normcore." It brought out the snark in me. "Everyone looks ordinary except not as good as ordinary. Schlumpy. What's up with that?"

"I guess it's our way of showing we don't care."

Fridge hosed, swiveling it a step at a time back to the hut: "Maybe it's a sign that the world really needs VR now!"

"Oh, you don't know. We're calling it Shallow Alto."

"It changed that much?"

"It just seems like everything interesting isn't here anymore. The Suicide Club is in the city, the *Whole Earth Catalog* moved to Marin—sigh—and Survival Research Lab doesn't even come around anymore. No one interesting can afford the rents anymore."

You might not know about these early Silicon Valley institutions. The Suicide Club was a punk urban adventure club that would do things like climb the Golden Gate Bridge illegally. It was one of the progenitors of Burning Man. That's where "Leave No Trace" comes from.*

Survival Research Labs staged walloping, genuinely dangerous, and giant performance art with equipment scrounged in Silicon Valley. Like a living, unsupervised guinea pig operating—for real—a tank with a thirty-foot flamethrower. You'd have to sign your life away to attend a show. All of these scenes would play roles in creating the first VR company, but I didn't know that yet.

"It doesn't matter if they left. We're going to do the most interesting stuff ever, and right here."

"You seem more driven."

"I hadn't noticed until you pointed it out."

On Track

I had been building up a sense of mission for years and it was finally becoming more focused. I would prod the gang to build machines to make social VR possible, and promote VR as a suitably intense source of fascination to compete with the mind-control games and foolishness that Norbert Wiener worried about. VR would be the alternative to AI.

If a high-level strategy was becoming clear, the ground-level tactical game was still vague. Should we try a startup? Try to cajole a university or big company into supporting us in a VR lab? Just earn enough money from games or whatever to make the stuff with no regard for any existing precedent?

All of us wondered what style of institution we were building, but none of us knew. Maybe a fusion of lefty and business ideals? A tech company

---

* This is the celebrated prime directive for attendees of the mostly unconstrained Burning Man Festival, which brings tens of thousands of people into a wild desert region of Nevada once a year to enact eccentric art and happenings.

Kevin Kelly visited VPL in the late 1980s and took these pictures of Ann's early concept drawings that were still pinned on the wall. Upper left: Very early concept drawing for the VPL EyePhone. Like every other team making VR headsets since, we underestimated the ultimate thickness that would be needed. Upper right: An EyePhone in use. Lower left: Children become Punch and Judy avatars. Lower right: A person inhabiting a chicken avatar uses virtual X-ray glass to look inside virtual objects. We used images like this chicken in presentations at places like the Defense Department, and yet they still wanted to work with us.

based on consensus decision making? Would that be a crazy idea? (2017 Jaron intervenes to scream, *Yes, that would have been crazy!*) But at the time everything seemed possible and everyone was idealistic and young enough to go for nights without sleep to get the latest demo running.

We became ever more obsessed with building VR projects during the course of 1983.

One thing that was clear was that we couldn't do everything. I struggled emotionally with the thought, but it became clear that the full-on phenotropic vision was a project for generations, not a few years. VR could be ready in time to run real-time 3-D computers when they started to work, however.

David had been experimenting with one kind of visual programming language at MIT called dataflow. I talked to Chuck and the gang, and we decided on an in-between course that would include a few of the internal tricks we'd developed, like the high-level incremental compiler architecture, but would opt for the already understood dataflow paradigm for VR software, as it would certainly be a match.* Dave had just finished his PhD and moved out to join us. (Present-day digital artists will probably be familiar with a design tool called MAX, which uses dataflow.) Body Electric was the name Chuck chose for our new VR control program.

We'd also need a 3-D design program. You couldn't just go out and buy a 3-D modeler. Young took that challenge on and started on a project that would eventually become Swivel 3-D.

We also spent a lot of time on the tracking problem, which is the subject of the next chapter.

---

* Sorry for all the terminology, nontechnical readers. Some of these terms are introduced in the appendix on phenotropic computing.

13. Six Degrees
(A Little About Sensors and VR Data)

The Eyes Must Wander*

When Tom first made gloves, they measured finger bends, but not where the hand was in space, or how it was tilted. (You need six numbers to describe the position and orientation of an object in three dimensions: x, y, z, roll, pitch, and yaw.)

Obviously we'd have to be able to know where your hand was and how it was tilted if we expected your avatar's hand to be able to pick up a virtual object. Devices that can tell where an object is in space are usually called *trackers*.

Trackers suitable for human motion were already on sale, though they cost a bundle. Strangely, the state of Vermont used to dominate the tracking industry, such as it was back in the 1980s. Four different tracking companies called a single Vermont valley home. Their customers used trackers in robots, industrial equipment, even flight simulators.

In those days, there was always an external device, a base station that served as the reference point for trackers. For instance, two of the classic Vermont tracking companies (Polhemus and Ascension) specialized in tracking using magnetic fields. There would be a big electromagnet in

* Yup, a shout-out to Diana Vreeland.

an impressive enclosure emitting a pulsing field, and then there would be little magnetic field sensors attached to a glove, and eventually to the headset.

There were plenty of other potential ways to accomplish tracking, aside from magnetic fields; lasers, radio waves, on and on. We put a lot of time into ambitious tracking schemes.

Why put a tracker on a headset? Remember the spy submarine? It must probe.

Recall the fundamental principle of vision in VR, which was stated earlier in the section titled "The Mirror Reveals": "In order for the visual aspect of VR to work, you have to calculate what your eyes should see in the virtual world as you look around. Your eyes wander and the VR computer must constantly, and as instantly as possible, calculate whatever graphic images they would see were the virtual world real. When you turn to look to the right, the virtual world must swivel to the left in compensation, to create the illusion it is stationary, outside of you and independent."

Twenty-fourth VR Definition: A cybernetic construction that measures the probing aspect of human perception so that it is canceled out.

The crucial point I must get across is that the quality of the visual display on its own is not the most important part of the quality of the visual experience of VR. What's much more important is the tracking.* How fast and how well does the visual imagery respond to head or eye motion?

---

* There are different kinds of tracking: Since the eyes are approximately spherical and approximately rotate on their centers, you can often get by only knowing only where the eyes *are*, not where they're looking. As long as you can present a wide enough virtual panorama around where the eyes are, they can look around and see virtual stuff properly. That's called *eye tracking*. In fact, the eyes are in fairly fixed positions within the head as they swivel, so you can sometimes get by with mere *head tracking*. In some kinds of VR displays you must know the direction in which the eyes are looking, not just their positions. That's called *gaze tracking*. (There's no end to tracking, just as there's never any end to measurement. Sometimes it's essential to track the focal distance of each eye, or how open the iris is.)

The Brain Integrates

Twenty-fifth VR Definition: A media technology for which measurement is more important than display.

A universal problem in sensing is that it is a process, so it takes time. If you become a professional in the VR world, you will be asked to use the term "latency" for delays within VR systems.

The primacy of latency was demonstrated dramatically in the early 1980s. A VR lab had sprouted at NASA Ames, run by Mike McGreevy before Scott Fisher arrived.

Mike tried an experiment. He built a black-and-white VR headset that had a resolution of only 100 by 100 pixels per eye. That was the highest resolution possible, given the available display technology. The core rendering was still in vector graphics; a camera looked at the vector images to drive the pixel-based display. At the time, the mere use of pixels in a headset was still novel. This was probably the first example outside of flight simulators.

While 100 by 100 is a plausible resolution for an icon, for a virtual world, it's preposterous. Remember, the image is spread out over much of what you can see. Therefore, each pixel might look about as big as a wall tile! And yet the effect was amazing.

My jaw dropped when I looked at a simple outline model of a proposed satellite through Mike's headset. It looked reasonable! You could discern details smaller than the pixels and get a feeling for the 3-D form of the odd object.

The secret was reasonably fast and accurate head tracking. The lower the latency, the better the visual experience! It was as if the resolution had been magically multiplied.

Visual experience is based on integrating all you've seen and anticipating what you'll see next. The brain sees more than the eyes do.*

---

* An astonishingly dramatic yet everyday example is the blind spot. Each of your eyes is blind within a sizable zone that is not too far from the center of the field of view because that is where the optic nerve attaches to the retina and blocks its sensing function. And yet you are not aware that your brain is filling in that hole.

As you moved your head in Mike's HMD,* your brain was sampling how the virtual world looked from each slightly different perspective, moment to moment. As long as the moments were accurate, meaning that the tracker was good, the perspectives were also accurate. That meant that the brain could combine the stream of low-resolution images into a more accurate internal experience of higher resolution.

To the brain, this was nothing special, just another day on the job. The human eye is amazing, but also a squishy, inconsistent, quirky sensor. The brain always sees better than it should be able to, given the nature of our eyes. It is just as happy to guess and cheat in order to make VR look better as it is to do the same for everyday reality.

In the 1980s it used to be incredibly hard to explain this basic quality of VR to newbies, even when they could try a VR demo! I practiced for years to find the language to convey this simple idea, which by now ought to be utterly commonplace.

Twenty-sixth VR Definition: A media technology that prioritizes stimulating the cognitive dynamics by which the world is perceived over accurately simulating an alternate environment.

A Moving Experience

If you were one of the early subjects in our eccentric lab at VPL, you would have initially experienced the transition to "believing" in the virtual world. This was called the "conversion moment."

As VR improved over the years, the conversion moment happened sooner and sooner after a person put on the headset. Around the turn of the century it stopped being a thing.†

---

* For "Head Mounted Display," which was an early term for a VR headset.

† This is one example of an important principle: Cheap chips make other parts do more.

   Along with machine vision, chips that can sense motion though inertia have been getting better and cheaper. Every portable device has an accelerometer these days. Combine the data from an accelerometer with data from cameras and it's possible to create an even faster, more accurate tracker. Moore's Law swallows everything.

   And there's more: Fast chips make it worthwhile to attempt to predict the future. The usual

Today, the sensation for most people is sudden shock and delight at the quality of the illusion instead of a gradual onset in which there's time to notice one's own perception shifting.

This might be an example of better technology not actually being better. There is no greater value than learning more about oneself, and the older, poorer VR equipment might have done a better job at exposing one's own process of perception.

But that's no way to think! Perhaps a VR designer will come up with a crafty "slow start" experience on modern VR equipment that highlights the conversion moment even better than the old equipment did.

At any rate, in the old days you would have experienced the conversion moment, but that would only have been the first step. Since our emphasis was on multiperson experiences, you would soon have been introduced to another person inside VR.

You'd have seen this other person as an early avatar: a smooth, cheerfully colored figure with a cartoonlike head, an almost featureless body, and nimble but strangely tubular hands. The visual quality of VR during this period was rather fuzzy.

The face was generic. Everyone had to share avatars. There were few people who could make effective avatars out of the meager available resources, and little variety was possible, for the same reason. The first avatar face in VR was designed by Ann Lasko, and she made it out of twenty polygons; an origami face.

And yet, despite the paucity of visual detail, the presence of a human being came through. The effect was eerie and startling. In the course of everyday life, you don't particularly notice the difference in your state of perception when you become engaged in contact with another person, but in these crude early VR systems, the difference was highlighted and quite dramatic; a bristling of the skin.

---

math involved is called a Kalman filter. Just as your brain (probably the cerebellum) can predict where your hand will need to be in order to catch a baseball that is still in motion, Kalman filters predict where the head is about to be positioned. More specialized algorithms can take advantage of the particular anatomy of the body and neck; your head can only move in particular ways, so there's no need to consider impossible head motion.

Furthermore, by the time you've rendered a 3-D scene, it might be a little out of date because 3-D graphics are still a lot of work, even for today's cheap chips; so high-performance VR setups will make last-microsecond adjustments on a simpler basis to make the images just a little more current. (The overall image might be shifted, tilted, and warped, for instance.)

Suddenly another person was present there, in those few polygons! You could feel them, the warmth of human presence.

What was going on? If you recorded the motion of a person and played that motion back to animate an avatar, it would be apparent to people inside the virtual world that the avatar wasn't inhabited by a real person at that moment. Things were dramatically different when you were interactively engaged with another person, avatar to avatar. You could usually even tell who it was.

These early experiments with avatars paralleled those of a long-standing scientific community that studies the perception of "biological motion." A canonical experiment in that community is to make a movie of a subject concealed by a black full-body covering, but with a few bright dots stuck on here and there. All the movie shows is spots moving around.

Something remarkable happens when test subjects then look at such movies. They can often recognize individuals, or perceive details about sex, mood, and other qualities in strangers—all from a few dots in motion.

There are still debates about exactly what people can and cannot perceive from videos of such moving dots. Whatever hypotheses one prefers regarding biological motion, avatars probably reveal even more, because they are interactive versions of these experiments.

The visceral realness of human presence within an avatar is the most dramatic sensation I have felt in VR. Interactivity is not just a feature or a quality of VR, but the natural empirical process at the core of experience. It is how we know life. It is life.

Twenty-seventh VR Definition: A medium in which interactive biological motion is emphasized.

This definition excludes most digital experiences, such as games played by typical controllers, because they don't convey the continuous motion of the body, only button presses. It includes the most engaging Kinect experiences, and even the most provocative multitouch designs. It is the new digital frontier.

Trackers are not only the critical fulcrums that allow VR displays to work at all, they also measure people so that they can turn into avatars for each other. Sensors are truly the core technologies of VR. VR is more a science of measurement than of synthesis.

Whenever we set up a classic VR system, the first order of business was setting up and calibrating the tracker. What we all wanted was to not have to do that ever again.

A student named Bob Bishop scoped out a plan to get rid of external tracking base stations or reference points using machine vision—as a PhD dissertation, where else but at the University of North Carolina at Chapel Hill. It worked, but still required a prepared environment with visual targets to serve as reference points.

No one developed a headset with a completely self-contained machine vision tracker that didn't require a prepared environment or base station before HoloLens.

Sea Legs, Virtual Legs

No amount of heroics is wasted when it comes to combating latency. You count the microseconds.*

Twenty-eighth VR Definition: The digital medium that fights the hardest against time.

If you feel nausea in VR, the problem is often related to tracking problems.

A story I never tire of telling: Once upon a time in the 1980s, when VR became a pop culture craze for the first time, I got a phone call from a certain Mr. Steven Spielberg, a film director. "You've got to bring VR demos down to L.A. to show the entertainment industry. Maybe we can make VR rides for theme parks, or who knows what."

"Are you serious? VR systems need these giant computers and all this fancy equipment. Just tell people to come up to Silicon Valley for demos. It's a short flight."

"This is Hollywood. People come to us."

"Silicon Valley will change *that*, just wait."

---

* VR starts to feel good when certain perceived latencies get down to around 7 or 8 milliseconds.

"Maybe so, but in the meantime, we'll pay you to come . . . a lot!"

"Um, okay . . ."

Thus was launched Reality on Wheels, a big ol' 18-wheeler loaded with millions of dollars of VR demos that traveled from Silicon Valley to Hollywood. (Similar demos can be had for mere hundreds of dollars today.) It parked for a week at a time at all the major studios.

When it finally parked at Universal, where Mr. Spielberg was making his films, he was concerned that we had previously visited the competition at Disney. "The mouse has teeth!" A warning hissed in my ear.

The classic studio head of Universal, Lew Wasserman, a Borscht Belt Onassis, watched as we brought eager volunteers into the truck to try out this exotic new experience. They were dazzled, of course.

I occasionally meet people who had demos back in those days. A few are still active in the VR world today. They went on to write movie and TV scripts based on VR, or became venture capitalists funding VR startups.

Lew pointed at me with hand upturned and curled his wizened index finger like a magic wand, beckoning me. I hopped over like the eager bunny I was in those days.

"Kid, are people going to throw up in this thing?"

I replied in hyperventilated allegro staccato. "Great question, Mr. Wasserman! We've been studying the problem. We now only see nausea in one out of hundreds of demos, and in the near future it will only be one in tens of thousands. And then eventually it will become even rarer. We have this issue under control."

He snarled at Spielberg, "Why are you bringing me a kid who doesn't know the first thing about the entertainment business?"

To me: "Kid, I want to read headlines about how my janitors are quitting because of the vomit!" To be fair, vomit-motivated resignation was a trope that helped promote movies like *Jaws* and *The Exorcist* back in the day.

"If that's what you want, Mr. Wasserman, no problem!"

Today we have greatly reduced the occurrences of nausea in VR, just as I promised.

But despite our advances in addressing simulator sickness, we haven't achieved perfection. Once in a while I'll find someone who feels ill just watching someone else getting a VR demo, even if the other person is experiencing no problems.

I have seen a few people get woozy just thinking about VR. You can't solve all problems related to subjective experience without becoming the thought police, so we will have to live with a touch of imperfection indefinitely.

If only I had Mr. Wasserman's talent as a tummler.

Virtual Realism Versus Virtual Idealism

When designing a VR device or experience, one criterion is—of course—to make it as effective an illusion as possible. But I have argued that delineating the edges of the illusion makes VR better.

So we have come upon an interesting and fundamental tension. One goal for VR must be to make the illusion as convincing as possible, for otherwise what are we even doing? But the best enjoyment of VR includes not *really* being totally convinced. Like when you go to a magic show.

An example of this tension played out in the introduction of Microsoft's Kinect. I was so utterly thrilled with Kinect!

In the late 1990s a research coalition I headed, the National Teleimmersion Initiative, had created the first interactive experiences with depth maps.* That meant that people and environments were sensed volumetrically, in real time, in full 3-D. In earlier 3-D interactions, such as with the DataGlove over Seattle, your body was sensed here and there at spots, but in this case your whole shape was continuously scanned in 3-D. The software represented you as a moving sculpture, not a marionette with just a few joints.

The notion that a consumer device could bring this type of interaction to masses of people only a little more than a decade after it had been lab exotica was invigorating.

The introduction of Kinect also provided an exceptionally clear example of the tension between virtual idealism and virtual realism. Microsoft shipped polished titles like Dance Central, a popular learn-to-dance experience, but did not expose the raw data, the interior, practical truth of what the device actually did.

The gap in exposure utterly fascinated people, and a response welled up. A cultural phenomenon of "Kinect Hacks" arose in which amateur

* http://www.scientificamerican.com/article/virtually-there/

Image from a Kinect Hack video.

programmers wrote their own software for Kinect and posted YouTube videos.

These videos were not polished. Not by a long shot. They were nerdy and crude. Most of them simply exposed the raw data. You'd see gritty 3-D dynamic digital models of ordinary people at home in T-shirts.

Did the exposé ruin the grandeur of the illusion? No! When the internal nature of Kinect was revealed, huge numbers of people went nuts over it. Seeing the raw interior made the device *more* fascinating!

There were probably no more than a few thousand Kinect Hackers, while the commercial unit sold in the tens of millions, earning it the title of fastest-selling consumer electronics device in history. So did Kinect Hacks matter, or were they just inconsequential foam on the great sea of the consumer market?

I think they mattered. Even though the number of hackers was small, their cultural visibility was high. The hackers explained the device and helped set the tone for how it was appreciated by millions.

Seeing the raw data coming out of the first generation of Kinect*—the noise, the glitches—was intoxicating in a peculiar, highly contemporary

* A second generation came out that produced much smoother and finer data.

way. People finally saw what the computer could see, and that in turn explicated a layer of the digital world in which they lived.

If the Kinect Hack movement were to be encapsulated as a speaking entity, it would say, "This is what the devices that look at you can see. You are now a little less blind than before to the new world being brought about by techies."

Kinect Hackers, and to a lesser extend those who viewed the videos, entered into a conversation that is steering our civilization: How will people modify the sensorimotor loop through which we know and affect our world? Digital culture is all about modifying that loop, and Kinect Hack videos mostly showed quirky twists.*

Make oneself into a puffy, inflated version of oneself. Make oneself transparent. Hackers turned themselves into monsters or controlled waves of Christmas lights with waves of the hand.

It is a canonical cultural event of our times: coming up with a twist on our causal connection to reality and demoing it with digital devices. VR demos are like humor, opening up one's thinking a little bit.

Twenty-ninth VR Definition: A cultural movement in which hackers manipulate gadgets to change the rules of causality and perception in demos.

HoloLens doesn't hide its raw data the way Kinect did at first. To this day, when I put on a HoloLens, I remain fascinated by that simplest experience of seeing the world being digitized as I look around, everything getting wrapped up in simulated hunter's nets.

I have seen this process thousands of times over the decades, but it still engrosses. It is the gears of our algorithmic world turning, not a representation, but the actual turning. It is concrete; it is liberating. The calculations are amazing—but not perfect. An occasional bump, a gap, a roughness.

A good scriptwriter never tries to make the hero perfect, and yet we technologists often make the rookie mistake of trying to present our technologies as pristine.

---

* https://www.youtube.com/watch?v=ho8KVOe_y08

This is a book about VR, but I must mention that the Platonic sizzle way of thinking about technology extends way beyond VR. When companies design big data algorithms to recommend who you should sleep with or what movie you should watch, they expect an uncurious, gullible variant of the human species. People will play along, willing to get by without seeing what data is really in play or how the algorithms actually work.

One of the things I like best about the culture of VR is that many of the same users who are ready to accept the brags and theatrics of other digital services at face value are curious and even demanding to see behind the curtain of VR tech.

That makes sense. VR data is a derivative of personal, point-of-view experience. It's immediate and sensible. When you see VR data, it has a flavor and you can understand it. VR makes people curious and there can be no more important function for a technology.

Thirtieth VR Definition: A technology in which internal data and algorithms are intelligible as transformations of real-time, point-of-view human experiences and thus inspire curiosity to look under the hood.

Time to resume looking under the hood of my life.

14. Found

Bite Before Hook

1984 turned out to be an eventful year.

Young was developing his 3-D design tool, Chuck was working on dynamics, Steve was working on user experience, Tom was building different kinds of trackers. We were excited about the reveal of the Macintosh from Apple, and managed to get a modest level of VR-related experience to sort-of work on the earliest version, though not in true 3-D, of course. Even though it had only just been introduced, and was supposed to be a deep secret beforehand, we'd actually been keeping up with the Mac's development as it was happening. Steve Jobs would occasionally piss off one of his engineers severely enough to cause fleeting rebellions, so you'd see wire-wrapped prototypes casually exposed, tied on the backs of motorcycles when people were visiting the hut.

Andy Herzfeld, who had written the Macintosh operating system, had just left Apple after the blowups there.* He came to the hut and we built a mind-blowing Mac-based demo. It merged our antilanguage approach to high-level programming with glove-based manipulation, but also with link following and other hypertext-like elements. Yet another old artifact

---

* Apple famously fired Steve Jobs, and the whole Mac team quit. Then Apple almost died until he came back, and then it became the world's most valuable company. This is why people like Mark Zuckerberg are shown such deference today.

lost to the ghosts of old platforms, I fear. I don't even remember what we called it. Man, Andy can code. He was one of the best I ever met. (By the way, Andy was *not* among those who would spill Apple secrets, and I'm not telling who was.)

Then a fluke of publicity intervened. One of my earlier visual programming language designs was on the cover of *Scientific American.*

This came about because Larry Tesler, a scientist at Xerox PARC, had seen my work. It is still almost inconceivable how generous so many people have been to me over the years. Larry was known as the inventor of the browser, meaning not just the thing that looks at Web pages, like Edge or Firefox, but the much more fundamental concept of a choice-based interface for exploring information structures at all. There was a time when such basic things had to be invented. Larry went on to run research at Apple and later at Amazon.

Anyway, as the issue was being prepared, I received a call from an editor at the magazine who asked me for my affiliation. Not only did I not have one, but I had by then adopted a dose of hacker attitude, and didn't want one. That turned out to be a problem.

"Sir, this is *Scientific American*. Our editorial guidelines clearly state that the author's affiliation will be listed in the index and at the head of the story." After a few rounds of absurd argument, I gave in and made something up.

"My affiliation is VPL Research."

The editor sounded relieved, as if an irritating stone had been magically disappeared from his shoe. "What does it stand for, Visual Programming Languages?"

"No, it's for Virtual Programming Languages."

Suddenly, and I don't know why, I said, "Oh, make it VPL Research, Inc." Maybe it would be a company someday. Who could know?

The issue was published, made a splash.

Alan Patricof, one of the Valley's pioneer venture capitalists, saw that "Inc." appended to the made-up institution in *Scientific American* and came by to visit our forlorn, funky corner of Palo Alto. He looked at the demos and said—and these truly were his words—"Young man, you need venture capital."

I responded, "But there's no company!"

"We'll fix that right away."

"Can you give me a couple of days to think about it?"

"No time for slowpokes in Silicon Valley."

"Right."*

Peanut Sauce Gallery

Fate was caught on a knife's edge and I was given just a moment when I could prod it to fall one way or the other. Should I dive in and start a Silicon Valley company?

There was no beaten path. No startup incubators, no young entrepreneur awards, no crowdfunding sites. Furthermore, I had not grown up in a topiary world where one's cousin is a lawyer who knows a banker. I knew none of the right people and was clueless.

Today's Silicon Valley looks wild and "emergent" from a distance but is actually rather structured and formal. It has become a narrowing

---

* Patricof was one of the people who didn't do well, ultimately, from VPL. I feel bad about it. What I hear is that he never invested in VR again.

world of prestige insiders who invest in startups and a few big companies and then decide whether to buy them. At the time, we were still making it up.

I was conflicted about who I wanted to be. A vague internal tiger was stirring; maybe I should be a hero CEO like Steve Jobs. Or, another part of me felt more authentic as the hacker who would forever ridicule that CEO. Hacker culture held that CEOs were either morons or brilliant assholes. No middle ground.*

At the Little Hunan, one of my very bearded hacker friends, who wasn't even part of our VR group, said, "You have to start a company and keep total control of it."

Said another, "Don't insult him. Why should he degrade himself by turning into a suit?"

I couldn't get these guys to shut up. "Oh god, people! We don't even know if there will be a company."

"You have to keep total control because otherwise the idiots on the board will feel like they're supposed to, you know, *do* something, and they're always idiot psychopaths." This guy was a recent refugee from Apple.

"It's a nice thought, but I don't think a hacker has ever started a company that amounted to much and was also able to keep control of it."

"Someday, someday it will happen!" A round of murmured cheers.

It took decades, but the dream of hacker control was finally realized in the new century. Facebook became the first giant "public" company unambiguously controlled by a single techie. But back to the Hunan.

I started to daydream with my mouth. "Maybe we don't need to start a company. What if it was just an art project? We'd make VR equipment, distribute it for free like the Free Print Shop up in the city."

"I don't know. Have you ever run an electronics factory? It's pretty hard to get it right. I don't see a volunteer group getting it together."

"And who is going to take support calls?"

Death blow!

It's hard to convey what support calls were like when digital goods were still novel. People would call and say, "I just bought Moondust for my

---

* If it was ever true, it is no longer true today. Silicon Valley is now graced by some remarkably brilliant, decidedly non-asshole CEOs.

kids. There's just a little plastic box inside the cardboard box. Where are all the sparkling lights and the music? Do you shake the box?"

"You plug that little plastic box into the matching slot on the computer."

"We bought a computer, too, but it doesn't seem to do anything when you plug it in."

"Did you connect it to the TV?"

"You connect computers to TVs?"

"Yes, yes, and then the sparkly stuff will appear on the TV."

"Can you connect the little plastic box directly to the TV?"

"No, the computer needs to be in between them."

Calls like this would come in without letup throughout working hours. People spent $40 or $50 on a Moondust cartridge and most of that money went into giving them basic computing instruction on the phone.

Everyone at the Little Hunan fell silent imagining what the support calls would be like for VR. Remember, at the time, there had not yet been a VR product, or even a VR experiment that looked much like what we think of as VR today. It was all in our heads.

"Oh god, you're right. This will have to be a company. No way anyone will take all those megamoronic calls unless they're paid money."

"Do you mean communism is doomed?"

"You know, I never thought so before, but yes, we'll have to get a lot of people to do incredibly boring things in the future if we want to have a lot of computers, and only capitalism can get people to accept being bored."

"According to Moore's Law there will be *billions* of computers by the end of this century. Where? In doorknobs? All those passwords! Can we grow the human population fast enough to keep up?"

"The only possibility is to get people to do their own support for free."

"Not possible."

"Sure it is. We'll use computers to train people to take care of computers. And they'll pay us even though they're doing the work." (This person eventually became an early employee at Google.)

"*Guys*! Stop. So far there's no company. There's no commune. Just give us five minutes to get something working."

The lot of us shut up and slurped dan dan noodles in silence.

Dotted Lines

The next constituency to confront was vastly more important and also more tender. I couldn't get it together to address the whole group at once, but brought it up with each person who was working with me on VR-to-be.

"Why should you be the CEO? You're kind of flaky."

"Yeah, I know, I've been thinking about that. Maybe I could be the CEO to keep idiots on a board from ruining the company, but we could also hire a president to be the responsible day-to-day person."

"I don't know, don't you think you need to really do it if you're going to do it?"

"Yeah, that's what I've been worried about, too."

"What about stock? How would you decide who got what?"

"If this happens, I should hold a majority of it for stability, but we could work out a way for people to get compensation in other ways. Maybe people who feel strongly about it could keep partial ownership of their projects, so that would make it more fair."

"Sounds complicated."

"I guess we can only dive in and treat it as an adventure."

"We're all young. I guess it won't kill us if this is stupid."

"Somehow I don't think that's what people are supposed to say when great companies are founded."

I called Mr. Patricof back, told him it would be a go. "Great. Put me in touch with your lawyers."

"Um, okay, I'll get back to you later today."

"You *do* have a lawyer, right?"

"Sure, just need to see which one should handle this."

Quick calls to GNFs and voilà, I was in the office of one of the prestige Valley lawyers that very afternoon.

It was impossibly glamorous to have a lawyer. I asked the guy what I should do if I were ever arrested, and awaited his response as if I had just said something impressive. I mean, I'd never been able to even think of affording my own lawyer back in New Mexico. Having one was like a seal proving that I'd arrived in fancy town. The guy was a business lawyer and found my question too weird to acknowledge.

"Patricof, eh? How did that happen? Who introduced you?"

"He just called me up out of the blue."

"Good start. Who do you have lined up to fill out the round?"*

Time to bluff. "Um. People call me all the time. I just expect the people I need to appear. I don't mean that in a New Age way; it's just that the word has gotten around about what we're doing and people call."

My bashful country boy innocence would fade, but at the time it was still a big blot all over my face.

"Hmmm. Might work. Would you mind if I pointed you to people who can help? Remember, I'm *your* lawyer. You don't need to impress me."

"Okay, right. Yes, it would be great if you could help, actually."

Documents were laid before me on the desk. Sign them and the first VR company would be conjured into existence.

I picked up the pen and time slowed. The pen in my hand slid though curvy paths, depositing quick-drying ink.

Fantastically bizarre, that these large bipedal mammals would be gently guiding smooth oblong objects around on fragile paper to make these tiny marks, and then treat them as significant.

Round Horizon

"Filling out the first round" turned out to be both easy and hard.

Easy, because we had killer demos and amazing people. When prospective investors came for their demos they would become so excited that they'd start vibrating. More than once I heard visitors exclaim, breathlessly, that they had encountered a "religious experience."

You have to remember that at the time our demos were unlike anything else you could experience. If you could go back in a time machine with today's base of expectations, I doubt you'd be all that impressed. It's all relative.

As for the dreaded "money stuff" so derided at the Little Hunan, I described a simple three-part business plan: (a) Develop high-end VR products to sell to corporate, military, and academic labs for millions of dollars per station; (b) spin off consumer products like VR gloves for gaming and 3-D design tools; and (c) create valuable patents to license IP.

---

* A startup must define successive rounds of stock with a specified number of shares, cost, and shareholder rights. Typically a round is sold out before a new round is introduced, and the earlier investors get better terms but higher risks.

We would thrive and grow, or in the worst case survive, on these "three legs of the stool." Eventually we'd get bought by a big company, or else would hold out long enough for VR to get cheap enough for consumers, and then we'd go public.

So far so good, but I pushed VPL to be experimental in too many ways at once. For instance, I stuck to the awkward plan of owning most of the stock, but at the same time remaining pure and not acting like an executive. There would be a president who would be the true suit, but without as much power.

Investors didn't like that, but they accepted it. It wasn't the only problem. The technical principals would maintain a degree of autonomy as a balance to my control of the stock; investors feared that would make it hard to move together as a team. Those investors were eventually proven correct.

In retrospect, everyone might have been served better if the early investors had been more assertive, but what could they do? After extended, awkward consideration, the round was filled.

Deputy Top Suit

Another challenge was to recruit a president so that I wouldn't be the only top suit. There were a few trial candidates who looked good on paper, but didn't really do much.

This still amazes me: There's a layer of people in the business world who are good at looking right for various high-end jobs, but not at doing them. One guy looked serious, but he spent all his time obsessing over just the right shade of blue-green for the company's brochures. I was incensed. "Hey, I'm supposed to be the creative lunatic. You're supposed to be the adult out there hiring people to make the manufacturing line work better."

I was learning how to be a CEO, but I remained crazy tender and naïve for the first year or so. The rosy glory of entering the legit world blinded me to the most basic rules of the game—at least at the start. I had the worst poker face in the whole geographical region.

Babe in the woods. One time I signed an important contract for VPL that I thought had been vetted by the lawyers without checking the backside of the paper, where the other party had snuck in additional language that ended up screwing us. No hard feelings. I worked with the guy again later.

You have to be graceful about the aggressiveness of tech business culture. Around 2013 I was at a big Silicon Valley wedding when one of the best-known and most classic venture capitalists, not Patricof but a party who invested in a later round, came up and cheerfully recalled how easy it had been to screw me back in the day. Fair game, I suppose. We laughed about it together.

Fortunately, one of the GNFs, Marie Spengler, from the Values and Lifestyle Program (VALS) at the Stanford Research Institute, came through with a president who would remain with the company for its duration, Jean-Jacques Grimaud. It turned out that SRI was helping a French startup that was trying to do something even more premature than VPL.

The Pocket Big Brain was the first device that looked like a smartphone. To be fair, it was an inch thick, had chunky pixels that only switched between two shades of gray, wasn't backlit, and worst of all, there wasn't yet a wireless signal for it, so it was a cry in the data wilderness. But the overall vision and design were implemented. It had a touchscreen, icons, a suite of apps, and a battery. The creators talked about a proposal for a wireless standard called 3G, which would someday bring data connections to the whole world, even outdoors. 3G did eventually happen, but only decades later.

The Pocket Big Brain was an even crazier project than VR, because we could at least sell expensive versions of VR right from the start to special customers. We had an immediate business. There was *nobody* ready to spend a million dollars for a pocket device that couldn't connect to a signal.

So Marie thought, "Why not see if the people doing this crazy thing might not be suitable for this other slightly less crazy thing?"

Jean-Jacques became the president and brought along a penumbra of euro investors, customers, and partners. VPL was suddenly an anomaly; a multinational startup.

Footprint

We moved into drab offices, just like any other startup outgrowing its garage. It was a difficult transition for me. I was still bathed in mood and had difficulty with the sterility.

After a short time I couldn't take it. We moved to a funky redwood building on the old marina in Redwood City. Most of our offices were on the second floor, right above the water, with sliding glass doors and a big shared balcony. There were Veeple, as we called ourselves, who lived on boats, and there was a cute deli with a dock. It was just great, so I hardly have to tell you that it's been demolished to make way for a mosaic of cookie-cutter high-end condos. So goes the Valley.

There wasn't yet a factory in China you could call to manufacture small batches of products to your specifications. We had actual factories in Silicon Valley, which produced chips and Apple computers and all the rest. That's probably the biggest difference between then and now. VPL had to create its own production line.

We started up a little factory in Redwood City to make headsets, gloves, and the rest. We did things that seem inconceivable today. We hired local people and trained them. Blue-collar local jobs! From a Silicon Valley startup! It happened!

But it wasn't perfect. There were ready consultants for every other aspect of being a startup, but manufacturing was still considered part of the parental, big-company world; part of the old economy Back East rather than the new Wild West. No one supported tiny-scale manufacturing in Silicon Valley. It was big or nothing. I wonder if America would have lost so much of its place in tech manufacturing if that divide had been bridged.

So that's my excuse as I convey that we never managed to produce hardware of a consistently high quality. It's the one thing about VPL that I still feel guilty about today.

I tried hard to source parts in the USA. A techie senator from Tennessee named Al Gore became interested and helped conduct a dragnet of American companies still making displays, but no luck. We ended up buying most of our parts from Japan. I used to go there so much; it wasn't unusual to have to fly to Tokyo and back twice within the same week.

We Shipped!

VPL seeded thousands of labs and businesses with the gear that allowed them to conduct basic VR research and prototype industrial VR applications. We often collaborated with our customers and pioneered VR apps.

The VR clothing items were expensive. The common EyePhone cost

over $10,000 back in the 1980s, and honestly was only barely adequate. The $50,000 HRX model was better, comparable to headsets that cost a few hundred dollars as this book is published.

We sold plenty of individual EyePhones and DataGloves, but our flagship product was a complete VR system, the RB2. That stood for "Reality Built for Two." You could actually unite more than two people, who became avatars to each other, but I liked the two-on-a-bicycle metaphor.

Alan Kay had called computers "bicycles for the mind," and in the case of VR, the metaphor was doubly informative. Tim Leary and a few early VR researchers were already imagining VR to be "electronic LSD," but actually it takes attention, effort, and skill to enjoy VR. Like a bicycle, not a roller coaster. Also, I always wanted to emphasize personal connection between people over superhuman noospherian* concepts. Maybe we'd

---

* Pronounced "no-oh-sphere." This used to be the preferred hacker term for the world brain super-organism posthuman artificial intelligence that would supposedly come about because of algo-rithms on the Internet. A noosphere might incorporate humans as cognitive elements or might function without humans. No one saw much distinction. The term was originally coined by Pierre Teilhard de Chardin in the 1920s as a way to focus thinking about the sphere of human thought. Hacker thinking today doesn't use the term as much, but it still embraces visions of a future global level of organization that would surpass earlier structures such as religions, markets, and nations.

collectively create a global virtual space, but even then, the connection with another individual should remain more cherished.

An RB2 was expensive; in the millions of dollars. The biggest and most expensive parts of the RB2 were the computers, usually from Silicon Graphics, as big as refrigerators.

The main difference between selling parts like EyePhones and Data-Gloves versus whole systems, like the RB2, was that customers buying parts often wrote their own software. They had their own conceptions of how VR should work, and we were happy to help them realize their visions.

When a customer ordered a whole system, though, VPL supplied the software. The software was really the heart of VPL, even though the hardware is better remembered. That is because hardware pieces become durable relics that can be photographed. EyePhones have been used as movie props, but the software remains difficult to even explain if you can't try it.

This is probably personal bias, but so far as I can tell, our VR development tools were better than the ones I'm aware of today. You could change everything about a virtual world while it was running, and you could do it either using visual programming or an interface that looked more traditional.

Our software wasn't perfect, and that was due to VPL's weird structure. (Company structure and the architecture of the software made by a company inevitably reflect each other.)

It's amazing how powerfully nerd allegiances swayed us and continue to today. Young wrote our 3-D design tool Swivel in a computer language called FORTH that had a certain insurgent appeal. Chuck did not use FORTH. Therefore the dynamics and geometry had to be adjusted in different programs. They could run at the same time, but an artificial divide in the concepts was enshrined. We never shipped a unified design that would have been so much better. Indeed, if we had shipped a unified system, that would probably be the standard today for everyone. We were there first and set the course. That's how things go in Silicon Valley. We cast long shadows, for better or worse.

While I wish they had been more reliable, I loved our products. I have an EyePhone and DataGlove by the desk where I'm writing and I still get a warm feeling from them.

15. Be Your Own Pyramidion (About Visual Displays for VR)

Remembering the EyePhone

The EyePhone was not only the first commercial VR headset, it was probably the first example anywhere that looked designed, that didn't have metal rails sticking out. It was also the first color, head-supported VR headset, even counting ones from research labs, so far as I know.

EyePhones were great! I can still remember the glow of anticipation I used to experience every time I was about to put one of them on. As objects, the earliest ones looked a little like the present-day Oculus Rift. They were black, had Velcro bands, and stuck out a fair amount. Subjectively the visual experience was most like the present-day Sony PlayStation VR headset. Eye-Phones revealed the virtual world with a similar diffused visual quality.

The worst problem with the early EyePhones was probably the weight.

Weight was a huge problem for the first half century of VR goggles. Ivan Sutherland nicknamed the support for his 1969 ur-HMD the Sword of Damocles because it had to be hung from the ceiling. There was a death, eventually, resulting from a cable failure in a different heavy HMD—part of an experimental 1970s military training system.

VPL's early EyePhones used a thick stereo magnifier* (from a boutique

---

* To bring the little display screens mounted in front of the eyes into focus and fill a wide field of view.

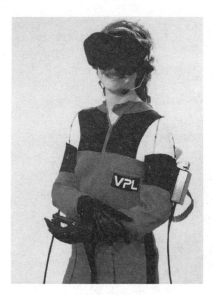

Ann in VR, from the outside.
Self-portrait.

optics company called LEEP) that could be hefted by a human neck, but it was indeed a heft. Another of the early VR companies, Fakespace, used the same optics, but their viewer was supported by a little crane.

In the 1980s you could cruise the sidewalk cafés on University Avenue in Palo Alto and tell who had gotten a demo at VPL in the past day or so. The pressure left characteristic red marks on the face. Tribal marks, we used to say.

In the late 1980s, VPL switched to lightweight Fresnel optics—those thin magnifiers composed of concentric circular ridges. VPL's Fresnel design was largely the work of Mike Teitel, one of the early Veeple.

We were able to achieve a resolution and field of view that would hold up today, though we had to charge fifty grand per pair, and that number isn't adjusted for inflation. I had missed those lighter goggles for many years, but lately the new wave of VR entrepreneurs have rediscovered the joy of lightweight optics.

That Thing on Your Head

The worst thing about big, classic VR goggles is also the best thing. VR headsets are the least fashionable fashion accessory. But I like that.

The obvious awkwardness is precisely what counters the potential for creepiness. There's no pretending you're not in VR when you know that from the outside you look like a psychedelic hockey player from a 1950s pulp science fiction illustration of sports on Mars. That's how VR *should* be.

Thirty-first VR Definition: You are having interesting experiences but look preposterously nerdy and dorky to onlookers.

Jaron inside VR as viewed from outside VR.

The desire to make VR equipment as close to invisible as possible has always seemed misguided to me. Consider Google's foray into heads-up displays, Google Glass. The more designers tried to make Glass blend in—just a tiny little fashionable thing on the face—the more it stood out. Like a pimple.

The question of what stands out in a design is always part of a negotiation about power. There's a conceit in Google Glass and related devices; the wearers of such devices will eventually be given the stealthy superpower of omniscient X-ray vision. But to an unadorned person nearby, it can feel like a surveillance device, as if the human face had been redesigned into an Orwellian demon mask.

But—and this is the core problem—both the wearer and the supposedly less empowered naked face observed by the device are in fact subservient. In the terms of information superiority, whoever is running the cloud computer that oversees the whole arrangement from afar is the master of both people. Even the wearer is worn.

So the pursuit of the fantasy of a superhero's magical psychic powers actually serves as a cover for submission. The tiny optics hanging by the eye make the whole face small.

As always, I am in a conflicted position, since some of the folks who propelled the Glass project are old friends.* I have also experimented with designs that resemble Google Glass, and if one of them had taken off, maybe I'd have found the rationalizations to love it. Only you, the reader, are in a position to judge my objectivity.

At any rate, here is a principle that is good and true: Bluntness is good in the design of information devices. Power relationships are unavoidable, but are always more ethical when they are stated clearly.

If a camera is looking at you, it should be visible. If the world you are wandering is not real, that should be made obvious. The human mind has a profound capacity to engage in fantasy, so we don't give up much if an illusion isn't perfect. At the same time, since fantasy comes to us so easily, it's usually good citizenship to emphasize the edges of an illusion.

The magician has a stage, which is distinct from the rest of the world. Were it not for the stage, or at least the announcement of a trick, magicians would be charlatans.

Maybe one's attitude about these questions has to do with how much one likes physical reality. I adore the natural world and love being alive. VR is part of a wonderful universe; neither a means of removal from it, nor a fantasy of getting the better of it.

I also like virtual reality very, very much, and that makes me even less interested in attempts to make it ambient or undetectable. I love classical music but am dismayed when I run across a person who leaves a feed of classical music on "for relaxation." It's so much more than wallpaper, if only you give it a chance. Less is often more, because attention isn't infinite.

When to Take a Pass

Another ethical crossroads in VR headset design will come up soon: There are two ways to implement mixed (augmented) reality. You can use optics to combine the real and virtual worlds, as we do in HoloLens. The images you see of the real world are then made of the same photons you would have seen from the real world if you weren't wearing the headset.

But there's another way to implement the effect, sometimes called "video pass-through." In that case, cameras facing out toward the world provide an

* At I write this book, the most publicized device in the genre is probably Snapchat's Spectacles.

image stream to conventional, or classical, VR headsets. Everything you see is then sourced from displays in the headsets, but does represent the real world. Video pass-through headsets are not unlike night vision goggles.

Possibilities are opened up when you use video pass-through. Your own hands and body can be modified, for instance. You can turn into a velociraptor.

The world can also change. Ran Gal, of Microsoft Research, made a filter for such headsets in which everything you see is transformed so as to still be functional, and have the same dimensions, but to be instantly redesigned as if from the set of the Starship *Enterprise*. It's fun and fascinating, and as research, Ran's work is fantastic.

Someday, society might progress enough for a consumer product along those lines to be ethical. But we're not there yet.

We've seen the damage to society wrought by a plague of sadistically false news.* A plague of sadistically false elements of reality would be both dangerous and present insane levels of potential for abuse of power. Control someone's reality and you control the person.

The Ridiculous Mistake That Floats in the Air All Around Us

There's a bit of business I've been putting off; a sad explanation of why most depictions of VR you have seen in science fiction movies, concept videos, and TV shows are physically impossible.

Again and again we see virtual stuff floating in midair. Princess Leia did it, of course, but the depiction is almost universal.

I don't mind it in science fiction, but it's also used in videos from defense contractors and by companies promoting VR products in a deceptive way. It's been used to cheat people out of money on crowd-sourcing sites.

Worse, all of this often happens in a *self*-deceptive way! I periodically run into military brass or tech executives who were so enchanted by the lure of floating holograms in videos—that they themselves commissioned—that they poured huge sums into a technology that can't actually exist, at least not in the present era.

It has been an expensive problem! I can casually count *billions* of dol-

---

* Appendix 3 examines this problem.

lars that have been misspent over the years due to people thinking that virtual stuff can be made to float in the real world in any arbitrary spot, rather than right in front of a special optical surface, or by wearing headsets or another intervention.

That can't happen!

Well, okay, I am certainly familiar with Arthur C. Clarke's famous edict that when an expert says something is impossible, he's almost always proven wrong, eventually. Maybe someday we'll be able to manipulate an incredibly intense artificial gravitational field that interacts with photons in order to deflect them with precision in a room, yet without tearing apart the flesh of human bystanders. It's not strictly impossible, perhaps, but it is utterly unimaginable within the world of options we inhabit today.

Here's why: Physicists understand photons pretty well by now. The quantum field theory that describes photons has as close to a perfect record as one can hope for at predicting their behavior in every experiment that's been tried.

One thing we know is that photons don't have any memory registers within them to tuck away instructions for a trajectory change that will take place in the future. Once they're traveling in a direction, they'll just keep on going until they interact with an object that deflects them.

That means you can't send a photon out into a room and have it make a preplanned right-angle turn toward your eye in order to be visible. There has to be a physical thing that you're looking at, or through, that's the last thing the photon touched before it hit your retina.

That last optical object could be a shining pixel within a screen that makes photons in the first place. That's the situation with a regular TV or computer screen. Or maybe a wavefront of photons bounced off a mirror. That's what happens when you see yourself brushing your teeth. Or maybe some photons slogged through the density of a glass lens and ended up heading in modified directions. That happens with normal eyeglasses, and it's called refraction. Or the photon might have been steered by a microscopic structure within a grating or a hologram. That's called diffraction.

But there can't be virtual stuff floating in the air in front of the mad scientist's glassware or the secret agent's gun cabinets, as seen by the naked eye.

I know this is a terribly disappointing revelation.*

Thirty-second VR Definition: The technology that is often misrepresented as being able to make so-called holograms float impossibly in the air.

You can sense my desperation. Why is this so hard for smart people like investors and military planners to understand? It's like trying to talk people out of bogus, expensive quack medicine. Human beings love to believe in the impossible.

A consolation is that there are a lot of ways to design displays for VR that do work. Whenever I think every possible VR strategy has been invented, someone comes up with a weird new idea. The possible turns out to be more interesting and fun than the impossible once you give it a chance.

---

* There are a few ways to slightly fake the impossible. You can heat air up with powerful lasers until it ionizes and thereby cause little bright bluish stars to spark into being in midair. A small number of such sparks can be coordinated and replenished often enough to form rudimentary floating 3-D phantasms, (This is just the flavor of extremist VR experiments that one expects from the energetic Japanese VR research community.) See http:www.Lashistar81;p/pdf/2016to6.pdf

Air isn't nothing; it bends light a bit. It is possible to coordinate intense sound waves to create dense pockets of air that bend light more than usual—but not enough to turn a photon at a hard angle from the middle of the room toward the eye. But maybe there's a way to at least make a cool demo. No one has yet made even a rudimentary demo of this approach to the impossible, to my knowledge, but sooner or later someone probably will. It will be outlandishly impractical.

The closest approach thus far to a "hologram" floating in free air was probably the one prototyped by my inventive friend Ken Perlin. Ken's rig scanned a small airspace for dust using invisible-light lasers and then immediately lit up fortuitous dust particles with bigger, visible-light lasers in order to create the effect. While the approach of making dust light up works a little bit, the result is inevitably quite fuzzy, dim, smeared, and pockmarked.

There are some other approximations: A bright projector can just project images onto whatever physical objects are already in a room. Some of my colleagues at Microsoft Research, especially Andy Wilson, have explored what can be done by coordinating projected images to match up with the real physical stuff that's already there. They can create the illusion that the room is pulsing, and other interesting effects. If people wear 3-D glasses, then 3-D images can be injected into the experience of the room, but that gets away from the "glasses-free" fantasy.

If you happen to have an interior decoration fetish for white, smooth-but-not-shiny surfaces, you can treat your whole room as a general projection surface. The effect can be useful in stage productions, and in some carefully planned interactive art scenarios. Michael Naimark was the pioneer of this approach, which is sometimes called "projected augmented reality," and there's a whole literature about it.

A Gadget Spectrum

Here's a chart that organizes many of VR's demonstrated optical methods (as opposed to impossible floating holograms) based on the location at which VR gadgets intervene in order to create the illusion of virtual stuff. I drew a simple chart like this when we started VPL to try to decide whether we should go for making headsets or another visual apparatus.

There are nine classes of VR displays shown here, and a total of seventeen ways to implement the visual side of VR, and this isn't a complete list! I realize this chart might intimidate nontechnical people, but there are really only a few important points you need to understand.

I enjoy working with near-eye displays the most (the familiar VR headsets such as the old EyePhones or the current HoloLens), so I put those options in boxes, but I have worked with almost every category of device shown here. One reason the list isn't complete is that my colleagues and I hope to add new entries to it, and I'm not ready to spill those beans just yet.

Why does this chart have to be so complicated? Why so many entries? The reason is that there isn't any one form of visual VR instrumentation that's the ultimate, perfect design. Each VR display is good at certain things

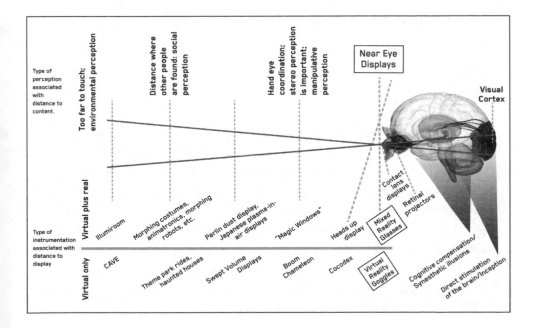

and not so good at other things. What I expect is that a variety of different VR gadgets will find places in our world.

VR is ultimately about people and our brains, so I organized the various ways of doing VR around a brain. From the point of view of a VR scientist, perception is organized in zones based on distance and location, with each zone emphasizing a different style of attention and perception.

For instance, the way you see the parts of reality you can manipulate with your hands is different from the way you see things that are too far away to touch. The nearer side of the divide is where stereovision is most important, for example.

We further distinguish what you're focused on, right in front of you, from what's off to the side, in the peripheral vision, which demands a different kind of vigilance. You are more sensitive to certain kinds of motion in the periphery, to the horizon, and even to slightly different colors, especially when it is dark. A well-designed VR headset takes all these fine points into consideration.

There's a big horizontal line in the chart. Underneath are found versions of VR in which you see only virtual stuff. Above the line is mixed reality, also known as augmented reality. In those cases you see the virtual intertwined with the real.

The Inner Extreme on the Spectrum

Let's look first at the option that falls the farthest to the right, because it illuminates one aspect of the practical philosophy of VR: It's been possible for quite a while to generate the perception of apparent light from electrical stimulation.

There have been preliminary experiments stimulating either the visual cortex or the optic nerve, as well as attempts to build artificial retinas. The results are still crude. The work has been framed as medical research, not media research, and usually a patient can only see a small number of dots. Most of the experiments are invasive. But progress is steady, so it's entirely credible to anticipate ever better artificial eyes for the blind in the future, should they want them.

Does that mean that VR ought to be accomplished by direct connection to the brain in the future? This is one of the most common questions I have been asked since the earliest days of VR.

It might be that direct brain stimulation will make sense in certain cases, but the question is misleading. It presumes that sense organs are dispensable, while in fact one would have to simulate them in order to simulate sensory experience. The brain and the sense organs are an organic whole. In the embryo they teach each other what form to take, and in childhood they train each other.

Remember, the eyes aren't USB cameras plugged into a Mr. Potato Head brain; they are the portals on a spy submarine exploring an unknown universe. Exploration is perception.

So the question of whether it would be better to bypass the eyes and go straight to the brain is misleading. The real question is, when would it be useful to simulate the presence of eyes, the way they look, probe, explore? The difference might sound academic, but it is crucial because eyes are part of the locus of power for the user who directs them; a direct feed suggests that power would reside instead with the source of the feed.

The Outer Extreme

Most of the optical strategies shown in the chart are positioned outside the eye and are organized by how far away from the eye they are to be used. Devices take on their characteristic forms based on that distance. To the extreme left, farthest from the eyes, VR displays become special rooms with amazing instrumentation. The canonical example of a VR room is the CAVE (Cave Automatic Virtual Environment), completely made of 3-D display walls. You do typically have to wear stereo/3-D glasses to use them, however. (In fiction, the corresponding fantasy design would be *Star Trek*'s Holodeck.)

CAVEs are great for experiences where your body doesn't become fantastical, and where the virtual stuff is too far away to touch. This category includes many scientific visualizations, where it's useful to be inside giant data sculptures. For instance, you might be inside a big model of a brain, watching the 3-D patterns of firing neurons. Or you might be floating above proposed megaconstructions in a city center.

CAVEs were invented by Carolina Cruz-Neira when she was a student of Dan Sandin and Tom DeFanti at the University of Illinois. These days she tends a wonderland: warehouses filled with different styles of experimental CAVEs at the University of Arkansas at Little Rock. Another example is UC Santa Barbara's Allosphere, a spherical CAVE with a cat-walk suspended in the center.*

My guess is that there will be a lot of VR in self-driving cars. It's almost intolerably boring to be in one, and we'll all be stuck in them for hours. The interior of a car is small enough to instrument without much trouble, but large enough to solve the duplex problem, which will be explained shortly. You can even cancel out the motion of the road to counter car sickness. VR and self-driving cars are a perfect match, even better than drive-time radio was with cars you had to drive. I wonder if people without property will spend a lot of time in VR, being driven from place to place because it's cheaper than standing still.

As VR displays get closer to the eyes, they get smaller. Beyond the reach of the human arm, but not as far away as the walls, a VR display might take on the size and shape of a large monitor or TV, except with 3-D and depth capacities.

In the diagram I use the term "Artificial Reality." This is in honor of Myron Krueger, who preferred the term and pioneered visual interaction on screens. His work is reflected today in highly interactive screen tech-nologies like Microsoft's Kinect, though there has yet to be a commercial realization of a complete VR screen.†

Continuing our traversal of the diagram, we'll for the moment pass over a couple of entries (morphing costumes and volumetric tank or dust dis-plays) to move closer still to the eyes. The hands might hold a gadget with the form factor‡ of a tablet (usually called a "magic window"). It would have to convey depth as well as stereo, and track eyes, just like its larger cousin. (A display that can do this would be called a lightfield display; a

---

* The Allosphere is down the hall from a branch of Microsoft Research. Station Q is integrated into the campus and is where mathematicians and physicists are trying to understand one kind of quantum computing.

† Here it's important to differentiate some distinct but similar-sounding gadgets. Everyone has seen 3-D TVs. A VR gadget in the form of a largish screen would have different capabilities than these TVs. You'd see depth, for one thing. 3-D TVs offer stereo, which has come to mean that each eye sees a different image. "Depth" means that, in addition, the eyes can focus, so that faraway things look fuzzy when nearby things look sharp and vice versa. But the more important distinction is eye tracking; the display knows where each of your eyes is and adjusts the perspective to match from moment to moment. (I explained why this is vital in an earlier chapter.) Most important, a VR big screen, like any VR display, must have a VRish input method. You don't just start videos in VR. You sculpt, fling, and glue things.

‡ A common Silicon Valley term for the size and shape of something. It used to be applied mostly to circuit boards but is now applied to any imaginable product.

slightly less ambitious approach that might also work well enough is called a multiview display.)

As with larger VR screens, no one has marketed a magic window as yet, though there are apps for regular tablets that approximate the effect. (David Levitt from VPL is offering one of them.)

That brings us back to the familiar VR headsets.

Scope Is the Thing with Feathers[*]

No VR headset is perfect, but the quest for the perfect headset has often turned out to be the driving factor in funding a VR project, because we care so much about vision. This is wrongheaded, since the other sensory modalities are no less important, A contributing factor is that engineers are drawn to headset optics because of its seductive landscape of engineering challenges.[†]

In my experience, when engineers first get into VR, they're usually obsessed with a solution to a subset of a list of optics/display challenges.

---

[*] If you don't get the joke, look it up in your Emily Dickinson.

[†] Obviously you can't just dangle little screens in front of the eyes, because they'd be out of focus. So at a minimum, one must bring them into focus, but that's not all. A partial list of other requirements:
- Field of view often becomes a macho contest. Who can design the widest field of view? Mark Bolas conducted the experiments that landed us with the benchmark of ninety degrees as a reasonable field of view for a consumer classical/occlusive VR headset.
- Images should not be distorted; stereo pairing should be correct throughout the field of view.
- In the real world, objects at different distances are focused on differentially by the eyes. It's nice when virtual stuff offers the eyes that option. This is called "accommodation" in the trade.
- Should be lightweight since necks can get cramped pretty easily.
- Center of gravity of head plus headset should be same as head without headset.
- Images should be sharp enough that you can read fine text.
- There should not be so much power flowing around the head as to be a hazard.
- Should not get hot.
- Should not get sweaty; there should not be condensation.
- Should ideally be able to operate without a cord, meaning on battery power and otherwise self-sufficient.
- Should offer contrast and gamut (range of colors) at least as good as the real world.
- Shouldn't flicker or have other disruptive artifacts.
- Texture, timing, distribution, and other qualities of pixels should either be imperceptible or pleasant.
- Inexpensive enough for practical use.

   The above list is only for classical, occlusive headsets where all you see is virtual stuff. Examples include the original VPL EyePhones and more recent products like the Oculus Rift or the HTC Vive. If we're talking about headsets for mixed reality, like HoloLens, then the wish list grows longer and the requirements shift. Mixed-reality headsets are much, much harder to design.

Teams get funded and build a whole VR system around a particular approach, certain that the rest of the problems will fall away, but so far that's never happened.

It's worth mentioning that optical designs for head-mounted displays usually start out on an optical bench. I adore this early phase of research. Lasers and mirrors are mounted on little metal posts on a special table that resists vibrations. These constructions always have a delicious mad scientist feeling, especially when the room lights are turned off and you can see the pure colors of laser light.

My colleague Joel Kollin, coinventor of retinal displays for VR, once suggested that we put a poster on the lab wall saying EVERYTHING LOOKS BETTER ON A BENCH. Getting from the bench to the head has been the downfall of the majority of high-risk or weird VR headset inventions.

There have been hundreds of optics/display designs for VR headsets, and each one solves for only part of the challenge. And yet it's absolutely hopeless to convince young VR engineers that they might eventually have to compromise, and to plan on that. They're always *shocked*, every single time!

Étendue

In practice, building an effective VR system is always a matter of balance and is always specialized to a purpose or setting. Even though the components have gotten much better in recent years, it will be quite a while before we can forgo trade-offs. That day might never come.

Thirty-third VR Definition: The ultimate media technology, meaning that it is perpetually premature.

The compromises and balances in each viable VR equipment design have their own charm. This is so in the same sense that it would be wrong to call black-and-white photography an obsolete form. It has its own culture, its own feel.

Perception is finite, like everything else. We can only highlight one thing by underplaying another. There is no perception without focus.

At each step of the way, each form of VR is a medium unto itself. These days we plow through them, discarding each design for the next, not taking the time to really know any of them. I suspect that future VR aficionados will retrace our steps, relishing every shade of difference.

The Duplex Problem

To conclude this chapter, I'll describe an unsolved problem in the design of VR headsets. VR remains a young discipline, still bountiful in mysteries.

A few of the experimental, one-of-a-kind EyePhones we built at VPL included sensors that faced inward at the face. Why? Remember, measurement is more important than display in VR, so any time you can measure more about a person, it will usually turn out to be important, even if an ultimate purpose isn't immediately known.

Maybe someday programmers will use facial expressions to adjust the finer points of algorithm design. Faces are expressive, after all.

That was the long-term goal, but the immediate motivation was to be able to embody avatars with expressive faces. When your real face smiled, your avatar's face would also smile.

In the 1980s this capability was too impractical to incorporate into a product. Optical sensors weren't good enough yet. We resorted to miniature contact rollers to measure how the skin was moving. (This was the era when mice had little roller balls instead of LEDs to move against the tabletop.)

These days, sensors are not the problem. With well-chosen optical sensors facing inward, it has become possible to measure not only where the eyes are looking, but changes in the pupils. Not only changes in the shapes of the eyelids, but in the transparency of skin around the eyes. Not only the shape of the mouth, but blushing. Cameras have certainly become adequately small, accurate, cheap, and low-power.

I've had a lot of fun with facial tracking. My favorite experience was when I was asked to keynote the NAMM conference, which is the big US trade show for the musical instrument industry. I triggered a set of sounds with a set of funny faces and practiced until I could get solid rhythms out of crazed repeated facial spasms. You can keep the joke running longer than any other funny thing I ever tried onstage. Why isn't this a big deal in pop culture yet? It would be so great in hip-hop. Yet another mystery.

Anyway, we can now measure what a face is doing, but when we place the appropriate sensors in a VR headset to drive an avatar's face, the result can be unappealing. We fall into what is famously known as the "uncanny valley."

The human brain is so finely tuned to watching the human face that if anything is slightly off, the strangeness quickly becomes creepy and alarming. It's called a valley because if things get *really* weird, if you're a lobster avatar, say, then your brain doesn't mind so much.

When the brain has good reason to expect to be well in tune with the world, then that trust must not be violated. When an avatar is weird but expressive, the brain is intrigued. When an avatar is just slightly off, then the brain panics.

You'd think that spasm music or lobster faces would be sufficient to motivate facial sensors in commercial VR headsets. Nope. Testers and focus groups always want to at least try being somewhat human, and then everyone gets disturbed because of the uncanny valley. Then, when the bean counters complain about how much a product is costing, facial sensors don't make the final cut. I've seen this process play out repeatedly, in various companies.

There might just be a gigantic payoff if we can bridge the uncanny valley, at least in VR headsets and avatars. It might make remote collaboration work better, and *that* might reduce humanity's carbon footprint. Transportation brings people together for meetings, classes, comedy clubs, etc., but burns a lot of carbon and causes a lot of congestion.

Straightforward camera-to-camera contact, such as in the familiar Skype experience, can do a lot, but not as much as we might like. Remember when I mentioned that there's a subconscious information channel between people that's transmitted by head motion? Add eye motion, skin tone, tiny changes in expression, and undoubtedly other factors we are not yet aware of. MIT's Sandy Pentland dubbed these the "honest signals." Without them, we perceive each other with less openness and ease, especially so between strangers.

People wear sunglasses to attempt to hide these signals, but it doesn't work, because sunglasses don't conceal head motion and other signal channels. The wearer can pretend that signals are hidden; it's a confidence booster, like makeup. That's all fine, but if the signals really are blocked, then people don't get along as well.

To perceive honest signals honestly, people have to experience each other accurately in 3-D. For instance, you need to be able to tell where an eye is to make eye contact. Everything has to be true to scale, even if the people are not physically in the same room. (Not that there must always be eye contact; it varies between cultures, but there's usually a world of meaning in the way a person *doesn't* make eye contact.)

It's not just about eye contact. The angle of observation is also crucial for perceiving skin tone, subconscious head motion, body language, even voice tone. This is another topic that deserves a whole book.

It's amazing to use a VR headset to look at an interlocutor who is being scanned in real time by an array of 3-D volumetric cameras. You can walk around and observe that person from any position, just as if you were in the same room. The transmission feels like a full-scale moving sculpture of the person, realistic; obviously not the physical person, but canny.

The effect was demonstrated for the first time in the 1990s, as part of the National Tele-immersion Initiative that I led. Recently, a Microsoft team led by Shahram Izadi showed a much better version, dubbed "Holoportation."

It's apparent to anyone who tries a demo of this experience that a product with the same capability would make it easier to establish trust, to keep meetings from wandering off-course, or participants from spacing out.

But. You are wearing a headset, so there isn't a way to set up two-way conversations. If the other person could see you, they'd see you in a headset.

The reason for this is that 3-D volumetric cameras have to be at least a little removed from the face of the other person to work like cameras. If you put them inside a VR headset, they get so close that you have to gather elemental data to reconstruct the face. Then you land back in the uncanny valley.

You are probably thinking to yourself, "How hard could this be to solve?" Couldn't you come up with rendering algorithms that bound over the uncanny valley? Or, could you make the headset transparent to the volumetric cameras? Or, could you make a headset big enough to give the cameras enough distance, but still be practical and desirable?

This is a great example of the kind of complex problem we work on in VR—at the boundary of cognitive science, cultural studies, sensor physics, advanced algorithms, industrial product design, and aesthetics. There have been dozens of partial solutions to the duplex problem in VR, but nothing ready to change the world.

Sounds easy to solve, right? Once it's solved, then that will be true.

Thirty-fourth VR Definition: Instrumentation that might just enable telecommunications with honest signals someday.

Back to Palo Alto in the 1980s.

16. The VPL Experience

In-Spiral

This is around the time when a normal memoir of a Silicon Valley startup would kick into gossip mode. Next would come juicy stories of intrigues on the board, struggles over stock, people yelling and quitting, backstabbings and betrayals.

VPL had all of that, and such stuff might make a good story, and yet it's not the story I'm telling. Here are a few reasons why.

First, you have to understand the most basic quality of the startup experience; you work harder than you had ever thought was possible. We worked so hard that there wasn't really a lot of time to reflect on what was going on as it happened. We all just struggled to swim.

A better metaphor might be falling into a black hole. You can't ever see a black hole because light can't escape. And yet astronomers observe them. How? Because as material is drawn into a black hole, it starts to spiral around, like water spiraling into a drain. On the way in there's so much activity to observe! That activity indicates the black hole, but it isn't the black hole.

The sheer density of work, once VPL got going, blotted out everything else. All of this book up to this point has been a spiraling into a period in which I didn't have the mental space to properly inscribe nonessential memories into my mind.

There's another reason. Remember that girl I picked up hitchhiking in New Mexico? The one who felt unburdened when no one knew where she was? Who thought of other people's attentions as invasive psychic tendrils?

Well, I'm having her experience, but as a writer. Most of the stories I've recounted so far haven't intruded much on other people's memories, but once VPL was founded, the stakes got way high for people other than me. Money and pride mattered to some of them, but it was also about identity and purpose—precious elements of life, not easy to come by.

I know there are still people around who cared tremendously about VPL, and I know they would tell different versions of the story.

Since I can't remember the densest times all that well, *and* I know that whatever construction I could come up with would trample tender concerns of other people, I wonder why I would even attempt it. Maybe I am choosing them over you, the reader. Maybe you deserve to read a juicy take on a bunch of old politics, even if it would upset someone else who lived it; or maybe even because of that. Maybe that's my duty as a writer. Maybe I'm not a real writer if I don't trample for your sake.

But then there's the third reason. The reason I don't remember the soap opera aspect of VPL all that well is that it is boring. What I could tell you would be similar to all the other stories of ambition and conflict you have read.

Screw that. The stories I can tell you about spiraling in are new to our age. You are getting the important stuff, and it's juicy in its own way.

Veeple

Many of the little stories and adventures one hears about in the latest wave of VR are eerily similar to old tales about VPL. In early 2015 an engineer at Valve* tweeted about the remarkable experience of falling asleep in VR and then waking up in it, which then motivated VR engineers everywhere to try to replicate the experience. Hackers often sleep in the lab anyway, after all. Yes, this had happened at VPL in the 1980s, first by accident and then by design. It's worth trying, by the way.

We were young, we were naughty. Margaret Minsky (Marvin's daughter)

---

* Valve is one of the companies that leapt into VR in the twenty-teens revival. It might be the most charming of the batch and reminds me the most of VPL days. The company is also known for the Steam gaming platform.

worked at VPL in the mid-1980s for a while on a project we cooked up for erotic wearables. Very Pleasurable Lingerie, it was called. The idea was that the lingerie would emit musical chords when touched, and chord progressions when stroked. The chord progressions would resolve to the tonic only in, um, certain places. I believe I saw this idea rediscovered on Kickstarter or similar recently. Hope whoever is doing it now completes the project. Very worth it.

There was also a vibrator terminating with a MIDI connection (the type of cable used to control music synthesizers) lying on a table in the reception room, apparently to perturb visitors. I'm not sure if it did anything, or who put it together. No one ever asked about it. I saw Brian Eno stare at it once, for a long time, but the perception didn't percolate through to a comment. Maybe he was watching us watch him.

I was the youngest person at VPL. Why? Was I still looking for a mother and surrounding myself with older people, if only a little older? It was hard to avoid taking on an adult persona even though I resisted. I wanted to be the rebellious weirdo, but I was surrounded by people who had more years of practice at doing that, so they outran me.

One time I got an urgent call from the State Department alerting me that one of our hackers had been caught smuggling marijuana in Japan. I'm not saying which one. This was a particularly gruesome predicament, as it could have meant lifetime imprisonment. I was terrified, but it turned out that the person in question quickly outsmarted the Japanese detectives and left them unable to prove their case. The whole affair turned out to be nothing more than high-stakes hacking recreation, immediately forgotten.

I can't name everyone who worked with us, but I'll mention a few other Veeple. There was Mitch Altman, whom we called Comet Mitch, because he was approximately seasonal. He would come around for half the year to help Tom with hardware; eventually he became one of the leading lights of the maker space movement.

Ann McCormick Piestrup. How do I even begin? A former nun, rosy and boisterous, a figure from a Manet canvas, Ann had become obsessed with the potential of computers in education. She started the Learning Company, which sold Warren Robinett's seminal programming game Rocky's Boots. It was the progenitor of builder games like Minecraft. Her hope was that we'd get VR tools to kids and transform teaching, math most of all.

Another remarkable programmer, a candidate for one of the best ever, was Bill Alessi. He had previously been at HP, where he was known as their resident code demon. He aspired to be a music star, and had the looks and talent to get there. He lived in one of the last remaining scruffy, colorful downtown Palo Alto hotels, a stand-in for NYC's Chelsea Hotel, where he had also lived. (No need to ask; it's been demolished.) He'd have to take a break from coding in the middle of the night, once in a while, to perform at a punk club in the city, but he'd always be back and his code was bug free.

So many more: Handsome George Zachary, who worked on the curious puzzle of marketing early VR and eventually became a prominent venture capitalist. Mike Teitel, yet another holographer, also from MIT, a sweet man, gentle, who designed later-generation EyePhone optics. Toward the end of my tenure, VPL got big enough that I no longer knew everyone. We outgrew our cute marina. (If you live in Silicon Valley, you know that tall building to the south of where the San Mateo Bridge alights on the peninsula? The one with the big octagonal window at the top? That was us.)

Mr. John Perry Barlow, who prides himself on having radar for fascinating women, would tell me about interesting women who worked there whom I never met. A single lass who looked like Audrey Hepburn and descended from Albert Camus? Maybe she was there, who knows?

I didn't just encounter new people, but new versions of myself at VPL. A laid-back country hippie morphed into a high-stress CEO. It's hard to believe this was me, but boy, I developed a temper.

After all my dismissals of superstitious thinking, I suppose I ought to report one example I can't explain. The engineers at VPL swore that when I got upset, I'd cause computers in my vicinity to crash, even through walls. Logs were kept, statistics analyzed.

Not only software, but hardware, too, was endangered, and not only through psychic forces. I remember one tense meeting with a supplier who wanted to be able to delay shipping us parts we needed, but without paying the contractual penalty. I stared menacingly at the company's representatives while gradually tearing a computer into tiny pieces with my bare hands as they watched. Not a word came from my mouth. We got our parts on time.

Afterward, the ever-polite Tom carefully gathered the shreds from the conference room table to salvage and catalog. I didn't always like who I was becoming.

What VR Was For

One of the most common questions I've been asked is, "What is VR's killer app? Is it just about gaming?"

The story of VR is still just beginning even as I write this book, so I still expect to be surprised by VR applications, but the apps we pioneered in the 1980s come up again and again. I suspect they might eventually be remembered as killers, or at least committers of misdemeanors.

I'll organize the apps we worked on at VPL by the type of partner who joined us. VR was all about partnerships. VPL was an instigator and accelerant, but was never alone.

(There were a few special partners that worked not on just a specific use of VR, but on a lot of different apps. Some were academic departments and some were startups. There are a few that stand out that were at once customers, collaborators, pods of fellow travelers, and coinventors: NASA,[*] the University of Washington,[†] UNC,[‡] and another startup called Fakespace.[§] It feels inadequate to barely mention people and places that were so important to the story, but at least I can give hints.)

---

[*] It was Margaret Minsky, naturally, who introduced me to Scott Fisher back at MIT. He was headed out west to be a researcher at NASA's Ames Research Center in Silicon Valley. His plan was to build a fine VR lab. Actually, Scott's preferred term was "virtual environment." His lab's work was iconic for the era. He made his own head-mounted display, and incorporated some of the first VPL gloves sold. Scott later started a department and taught at USC.

[†] Tom Furness was another formative figure in VR. He had been working on VR-like technology at the air force—simulators, heads-up displays, and the like—and had decided to switch to the university life. He founded one of the great labs, the HITLab at the University of Washington in Seattle. This lab had a particularly congenial atmosphere, and varied collaborations formed between it and VPL. The virtual Seattle that opened this book, in which my hand was enormous, was authored at the HITLab, though the bug in the hand size was not their fault.

[‡] I've been privileged to receive various awards and honors, but my most thrilling honor ever was seeing VPL equipment in use at the UNC lab. Exactly what I had hoped would happen was happening. By making the basic tools available, academic research was able to accelerate.

[§] Other little VR companies appeared in the early 1990s. They were often both partners and competitors, though no one else was crazy enough to manufacture and sell whole VR systems until years later. My favorite was Fakespace, which was started by Mark Bolas and Ian McDowall. They made a mini-crane-mounted VR headset that had properties a little like the EyePhone, and also partnered and contracted with interesting clients in the way VPL did, sometimes in collaboration with us.

Mark went on to become a professor at USC and played an outsized role in the VR revival of the twenty-teens. He designed an open source cardboard smartphone holder called FOV2GO that turned a phone into a rudimentary VR headset (under a Microsoft grant, by the way) years before Google issued its own version. That device made VR affordable and accessible to a lot of people for the first time. Mark also had his students design more substantial headsets, and a few of them went on to found Oculus.

The indomitable Sally Rosenthal using NASA's 1980s Virtual Environment system with VPL Gloves. The headset and overall system were designed by the VR pioneer Scott Fisher and his team.

Let's explore some of the other, more specialized partner/customers in order to understand the question of why anyone spent millions of dollars on VR setups in the 1980s.

SURGICAL TRAINING

Joe Rosen is a reconstructive plastic surgeon who worked at Stanford back in VPL days and is now at Dartmouth. He had once trained as a sculptor and has a wonderful feel for the body. He felt at home in the art community and famously put Mark Pauline's hand back together after an explosion, Mark being the driving force behind the notorious Survival Research Labs, the outfit with the superpowered guinea pig.

Joe and I initially collaborated on his "nerve chip," the first prosthetic nerve. When nerve bundles are severed and rejoined, they heal with the wrong mapping. Individual nerves connect to the wrong destinations. So while the bundle heals overall, it still takes the brain years to learn to deal with the scrambling. The plan was to put a silicon chip drilled with holes in the path of the healing of the nerve bundle and then have the chip remap the nerves properly. But how could we find the right connections?

The scenario we considered was that a patient with a severed hand would have it reattached with the nerve bundle healing through a nerve chip. (Unfortunately, patients with severed hands showed up in Joe's operating room fairly regularly.) Then the patient would don a DataGlove. When the patient tried to flex the hand or make a fist, the glove would detect in detail what was actually happening, and algorithms would adaptively remap nerve signals in the chip until the hand started to do what the patient intended.

This work was ahead of its time, and it didn't get very far, though Joe made the chips and demonstrated them in principle.

Dr. Joe Rosen with Ann Lasko. Joe is about to try our top-of-the-line VR headset, the VPL EyePhone HRX.

A little later, Joe, Ann Lasko, and I made the first real-time surgical simulator. A virtual knee. This work was eventually spun off into another startup that eventually morphed into a so-called medical informatics company that changed hands for billions of dollars—ending up as part of Pfizer. But that was long after my involvement.

The first surgical simulator was more a proof of concept. The second one was a little more challenging, this time a gall bladder procedure. Our medical collaborator was Col. Rick Satava, an army doctor who went on to start a medical VR research initiative in DARPA that was extremely influential.

Of all the virtual worlds I've worked on, the surgical simulators were the most satisfying.

Thirty-fifth VR Definition: Training simulators for
anything, not just flight.

TRADEWINDS

There was a particular bond between Japan and the early culture of VR. Scott Fisher adored going there, especially. In part, this was because Japanese culture felt exotic and symbolized the strangeness of the new world we were discovering in our labs. Walking in the blazing night of Shinjuku felt like what a virtual world might be someday. There was a lot of Japanese flavor in early cyberpunk, especially from William Gibson and in *Blade Runner*.

But also, the Japanese adored virtual reality. There were great early VR labs all over the country, and it was amazing to visit them. Henry Fuchs and I used to have a classification scheme for VR research—single person or multiperson, augmented or not, haptic or not, and so on—and we could find no way around having a separate category for "strange experiments from Japan." They kept on coming up with the most peculiar projects.

One time I was in Kyoto giving a talk about VR and I made a joke about how difficult it would be to generate virtual food. "The actuator would be too disgusting to contemplate. It would be an icky mushy robotic protrusion you'd stick in your mouth. It would simulate the textures of different foods and release tasty chemicals while you chewed on it."

A year later I got a note from a certain academic lab in Japan, one of VPL's customers. "We are pleased to announce that we have recently achieved the ability to disgust you." Sure enough, they had built a prototype of the revolting device, which required three forms of sterilization between each demo. I wonder where it is now. The thing ought to at least be used in a music video.

VPL had a Tokyo showroom, and the visitors were the most interesting people from Japanese cultural and technical circles. We used to be on Japanese TV a lot. I always felt a little embarrassed that our manufacturing quality wasn't up to Japanese standards. We were so sloppy compared to them.

One of the lucrative early VR apps was, to my surprise, a Japanese kitchen design tool. This was a collaboration with the giant industrial conglomerate Matsushita, and the VR experience was set up in a high-end kitchen showroom in Tokyo. They'd send a team to digitize an existing kitchen, and customers could come in later to experience remodeling possibilities in VR.

The hardest problem was modifying EyePhones so that they wouldn't mess up the expensive hairdos of women who came to experience unspeakably expensive potential new virtual kitchens. This hairdo problem had simply been moot for all the VPL engineers. The kitchen design simulator was profitable and stayed open for years, until VPL slumped and was no longer able to support it.

Thirty-sixth VR Definition: A way to try out proposed
changes to the real world before you commit.

We also had a distributor with a swank showroom in Paris, right across the Seine from the Eiffel Tower. A perfectly dressed young man—a model from Milan—was suspended in a glass box high above visiting customers, typing at an unplugged Macintosh at all times, for show. I have trouble making sense of the French, to this day.

Through our French connection we ended up working with the oil exploration technology company Schlumberger. One of the Schlumberger family kids even worked at VPL for a while. We developed early data fusion geo-visualizations. You could fly around in an oil deposit and model different drilling strategies. Commonplace today, but shocking and novel at the time.

Our customers included cities. We helped Singapore plan for its spectacular growth by building a model inspired by the virtual Seattle described in the preface. We helped the city of Berlin plan restorations after the wall fell, in collaboration with wonderful research teams from German universities and our German partner ART+COM. These renderings of Berlin were, I believe, the first virtual worlds with real-time shadows and reflections. We later reused a Berlin subway model as the setting for a scary virtual world for Universal in which train-sized snakes prowled and struck.

Thirty-seventh VR Definition: Instrumentation to
present data as lucidly as possible.

A FEW OF OUR AMERICAN PROJECTS

We helped Boeing build simulators for cabin design, field maintenance, and manufacturing line design. Boeing later became a key early driver of mixed reality, or "augmented reality," as they called it.

We also helped Ford and other car manufacturers use VR to prototype their designs, a practice that has long since become universal in the industry. And we did the same with companies designing trains and ships. Transportation was our biggest sector in a typical year. Every commercial vehicle you've entered in the last couple of decades was prototyped in VR. It's the quiet killer app. A serial killer?

One of our customers was a pharmaceutical company with a big secret. They planned to introduce a drug called Prozac, the first "blockbuster" antidepressant.

We were engaged to create a virtual world that would teach psychiatrists how Prozac worked. Upon donning the EyePhone, one sat in a simulated consultation room, and on the couch there reclined a simulated depressed patient. The rendering of humans required the absolute maximum of computer graphics power available on the planet at the time, but our client could afford it. We succeeded in making a character that looked depressed. I was quite proud.

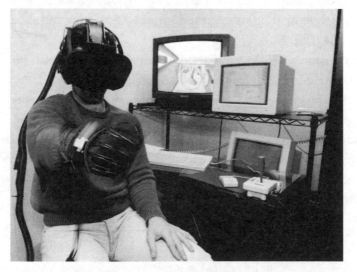

George Zachary of VPL trying out a driving simulator.

After a little patter, the psychiatrist getting the demo was shrunk Fantastic Voyage–small, and flew through the iris of the patient, up the optic nerve, into the brain. Then he got smaller still. We flew him up to a synapse; he could grab a Prozac molecule and shove it into a receptor in order to play with the chemical action. It was probably the most ambitious virtual world of the day, incorporating chemistry simulation and other hard-to-do things in one experience.

What I had not foreseen is that I'd spend exhausting, bizarre days at the annual psychiatrists' convention putting the world's top shrinks through demos in which they shrank. The situation was more surreal than even VR ought to be; in those days half of them looked like Freud impostors.

Thirty-eighth VR Definition: The ultimate way to capture someone inside an advertisement. Let's hope it is done as little as possible.

I wondered if playing with our engaging little promo world might be therapeutic for depressed patients. It turned out that VR was eventually used to treat depression.

SOLDIERS AND SPOOKS

At first I was wary of military contracts, and for good reasons. One was that most of us Veeple thought we might be pacifists. But also I was worried that military contracting companies tended to get locked into a cycle that didn't allow for much creativity. Contracting was all about meeting goals that were negotiated in advance. VR was so nascent that we *never* knew what our goals were in advance. Everything was a wild leap of faith.

Despite qualms, we learned to act as if we knew what we were doing, and signed on to a few contracts with DARPA and other military entities, creating wild stuff, unprecedented at the time, but I still can't describe the work.

As I met more people in the military I grew to respect them more and more. There were extraordinarily generous and smart folks in that world.

But I could also see how high-tech tools hypnotized military minds. I worried that we might be making our military *less* competent instead of more competent, but I didn't really know how to articulate my concerns in a way that would get through. It's still hard to convey such reservations to people who love the technology. Whatever one's feelings about pacifism, no one wants a less competent military. I still worry that high-tech tools can be a little too dazzling.

Some of the projects concerned visualizing complicated data in VR to make it easier to understand. I won't say what the data *was*.

If you turn complicated data into a virtual place, a palace you can roam or a city you can tour, your brain remembers better and notices more. Before printing, cultures all over the world had evolved "arts of memory" in which people imagined palaces or other places in which to position memories. In Europe they were called memory palaces.* You'd mount notices of facts to remember in pretty frames on your make-believe palace walls. Native Australians developed what might be the most elaborate example, called songlines. Your brain is optimized for remembering terrain. When we turn complexity into territory, we tame it.

Thirty-ninth VR Definition: Digital implementation of memory palaces.

This method was also applied to help veterans who had suffered memory impairments, so that they could take on new memories with a surer footing.†

Fortieth VR Definition: A generalized tool for cognitive enhancement.

Through the military, we met people in civilian law enforcement. We built a tool with the FBI to help them figure out where snipers might be able to locate themselves in order to threaten a public event. The hardest part, by far, was getting an accurate model of a city into the computer at that early date. We relied on surveyors.

---

* Sherlock Holmes has been known to use them as well, at least in his Cumberbatch incarnation.

† Conspicuously underrepresented in this list of VR apps is help for clients with disabilities. We actually did a lot with gloves for sign language, therapy for aphasia patients, and so on, but I have grown tired of the PR hype cycle around VR and disabilities, so lately I prefer to just act and not talk so much about it. The hype comes so easily, like a drug, that it can actually be an impediment to funders and organizations following through well enough to make a difference.

Forty-first VR Definition: A training simulator for
Information Age warfare.

While the app was deemed a success, it also unveiled dark possibilities. A different agency saw the demo and asked me if it could be used to choose camera placements in order to maintain constant surveillance of an individual walking about in a city with a minimum number of cameras. Um, sure.

A little while later the same question, roles reversed: Would our own spies still be able to function in the future, if there were cameras positioned to follow them without fail in a foreign city? How would a spy not be tracked?

My advice to our client was to hack into the foreign network and create the illusion of thousands of spies walking about, so that it would take time to sort out which signal was authentic. Old-school diversion to stay a step ahead. The strategy is routine these days.

(Later on, I came up with an idea for a scene in the movie *Minority Report* in which the hero is running from the police, but his image is incorporated into every advertising billboard he passes, so they can watch him without even trying. Everyone can. Indeed I implemented a model of the tech for a script meeting.) Our stealthy clients were delighted with my suggestion. A successful gig. As I was leaving, though, I suddenly had one of those chilling moments of doubt, like when you're hiking and you suddenly realize your step is not so sure and you're right next to a deep crevice. Take a moment. Think.

If a digital network could be used to hide the truth, and do that efficiently, right smack in the middle of torrents of information and openness, then why were we so sure that networks would serve the cause of truth overall?

But back to happy stories of early VR apps.

CHARACTERS

We might have been the first vendor of motion capture suits—our DataSuits—and we sold them to various people in the entertainment business. This was long before anyone could render a realistic CGI character for the movies, but there were still uses.

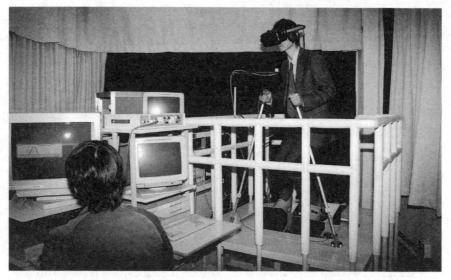

Skiing simulator using an EyePhone HRX.

For instance, there was a short-lived TV game show in which contestants wore DataSuits to control stick figure characters that had to achieve something—I don't remember what—but the general idea is worth trying again.

We even had a project with the Olympics to attempt to create a new sport in VR, which was way premature at the time, but once again, worth a reexamination.

We built theme park prototypes, mostly on Universal's dime, but none ready to be deployed. Our main collaborator was a film director named Alex Singer. He went on to direct *Star Trek TNG* episodes with their Holodeck. I loved visiting him on set. It might have been the last time science fiction was equal parts inventive, humanistic, and optimistic.

VR wasn't ready for the public in a theme park setting until much later, when Randy Pausch* worked with Disney. Randy was around during VPL times, though, as a professor starting out at the University of Virginia. We shared a conspiratorial belief that VR would become a new kind of language. (You've probably heard of Randy, not for his work in VR, but

---

* As long as I'm mentioning Randy, I have to also mention his PhD adviser, Andy van Dam. Andy, based at Brown University, is the master teacher of computer science. His students invented our age. Several have come up already in the book, like Andy Hertzfeld, but they're everywhere.

because of his famous Last Lecture, about living and dying well. Randy and I were about the same age, but Randy died, as close to a secular saint as one can get, in 2008 of pancreatic cancer.)*

There were a few people experimenting with adapting VR tech for the theater. George Coates came up with a tilted stage sectioned by layers of scrims that created the illusion that live actors were walking around virtual worlds and interacting with virtual stuff. He installed it in a cathedral-like space hidden within an early San Francisco skyscraper. The result was mesmerizing. People from VPL, NASA, and Silicon Graphics would sneak equipment to him and help out with the programming, though our digital stuff often failed during a show.

Jerry Garcia's[†] daughter Annabelle used a DataGlove and a skeletal hand derived from our research on surgical simulation to project a giant skeletal hand during Grateful Dead concerts. She said she liked to watch the eyes of fans follow the skeletal hand, all gathered into one gaze, like kittens watching a pendulum.

We collaborated with Jim Henson[‡] to prototype a simple computer graphics muppet, named Waldo. (Waldo eventually got a makeover and appeared as a fancier fellow than he had in prototype.) It was wonderful having puppeteers in our lab. I loved visiting the boisterous, messy Henson workshop in NYC. We learned so much from the muppeteers about character and expression, and they learned from us about the bizarre new idea that we had to design avatars without a particular camera point of view in mind. Jim was super lovely.

Forty-second VR Definition: Digital puppetry.

---

* www.cmu.edu/randystecture/book

† For readers too young to remember, Jerry was sort of the leader of the band the Grateful Dead, though the category of leader was antithetical to the very idea of that band. It amazes me how memory has disappeared now that the Internet is here. When I was growing up, I was aware of the music stars from previous generations, like Eva Tanguay, the vaudeville-era prototype for stars like Lady Gaga. Today, millennials I work with have often not heard of the Grateful Dead, even though it sometimes felt like an even bigger deal than computers in the Silicon Valley of the 1980s and 1990s. The Dead were associated with psychedelic drugs, a seemingly telepathic connection to their audience, and a following so passionate that many people's lives were fashioned around traveling to shows as the band toured. (I wasn't a huge fan myself, but I was an outlier.)

‡ Celebrated puppeteer; godfather of Kermit, Miss Piggy, et al.

Jaron teaching sitar to one of Jim Henson's Muppets. Also pictured is Muppeteer Dave Goelz.

We had a few mind-bending customers. Once I was flown in a private jet in the middle of winter to meet with the elders of the Sault Ste. Marie Ojibway tribe in Canada to evaluate whether virtual reality could be used to preserve their language, which was based on metaphorical references to events in tribal myths and was therefore poorly suited to the dictionary model of preservation. (The encounter indirectly inspired a *Star Trek TNG* episode about an alien race with a similar problem.)

My greatest satisfaction from VPL was visiting customer/collaborators. Whatever else happened, VPL succeeded in its core mission, to instigate and accelerate VR.

SPIN-OFFS

There isn't much in the above list about consumer-level VR, for the simple reason that VR wasn't cheap enough for consumers. But we did a few things.

The best known was probably the Power Glove, already discussed.

We also built wonderful prototypes of consumer experiences that never shipped, and will probably never be seen again. One used the old Amiga computer with 3-D glasses (the kind used for 3-D movies or TV, which are *much* simpler and cheaper than EyePhones) and Power Gloves;

an experience that was halfway between pinball and racquetball. But the Amiga didn't catch on, and it turned out that no other home computer that would work—color, 16-bit—arrived in time to revive the project.

We created toy prototypes. A teddy bear mirrored by an avatar in VR was named Nostrildamus, partly because the snout had sensors but also because VPL had an early logo called the Nose (the V looked like an upward-facing eye in profile, the P like an ear's pinnae, and the L like a fist with thumb extended).

Young's 3-D modeler turned into a stand-alone product, the first 3-D design tool on the Mac, called Swivel 3-D. It was spun off into a company called Paracomp that merged with another company, Macromind, that had the first animation editor for the Mac, and the result, called Macromedia, was eventually bought by Adobe, so a little of VPL lives on there. Swivel is still my favorite 3-D design tool, though it doesn't run on any current computer.

The investors were promised patents, so we filed patents. VPL patents were always a contentious issue. For one thing, nascent hacker idealism held the idea of intellectual property in contempt. For another, we arrived at the scene early enough to be able to disclose a lot of basic VR in patent form. No one had yet described how to join multiple people in the same world, link avatars to human motion, or pick up a virtual object with your hand as if it were real.

My friends in the hacker world didn't want us to file for these concepts, while the investors wanted us to file as aggressively as possible. We came up with an interesting middle course.

We did file for the patents, but we included the complete source code for everything, in total detail, thus nullifying trade secrets. On the one hand, whoever owned the patents could mine the code for further claims. Later on, when Sun Microsystems acquired VPL, that's exactly what happened.

But on the other hand, we taught how we did what we did completely. That meant that anyone else could know everything they needed if they wanted to work around our intellectual property. We were both open source and IP-driven, sort of.

Did that work out well? Not at the time. The patents were perceived as so valuable that people fought over them; probably ruined opportunities through an excess of conflict.

The VPL patents have all since run out. Ancient history.

17. Inside-out Spheres
(A Little About VR "Video" and Sound)

VideoSphere

We've almost completed a tour of the major parts of a 1980s "classical" VR system, and my entirely mild thoughts about them. There are only two remaining components: a kind of camera that works with VR and a way to make 3-D sound.

Graham Smith is the genial Canadian who first cracked the puzzle of generating encompassing, "spherical"* videos that could be enjoyed later through VR headsets. Graham had made his own HMD before working with VPL, which was no small achievement. He designed our environmental video capture and playback product, which was called VideoSphere.

VideoSphere was another VPL product that was ahead of its time. It was a strange-looking camera that captured a scene in all directions at once, though the geometry was a little trickier than a sphere.

This type of camera is not unusual today. You take a spherical video, perhaps at a concert or in a hyperactive urban environment; then a user can wear a VR headset to look all around as the video plays back.

I must digress on the topic of video in VR.

The limitation of spherical video capture in the raw is that it isn't

---

* The actual scheme is more complicated than taking literally spherical videos, but that's a reasonable approximation.

interactive. It's fun to have the perspective of being onstage at a big concert and being able to look around, but you can't *do* anything. You aren't really fully there, only a ghost. I ranted about that earlier.

But Videospheric recordings can be enhanced. You can overlay virtual stuff; computer graphics characters can be walking around in the captured real space, and they are fully interactive. They'll respond to you so you can feel present.

Computers have become powerful enough that we can interactively modify what goes on in spatially captured video. Video capture is no longer necessarily a record of reality.

It's generally easier to modify a spatial video than a traditional flat one, since the algorithms have more reliable hooks in the data to work with. If there's an old-fashioned two-dimensional video of, say, a police shooting, it's not that hard to detect if the video has been modified. A detail will be out of place, or a seam will show. But a spatial, immersive video record might be more easily modified in a way that tracks every seam and foresees every potential mistake. If the algorithms can access the full shape of a hand and a gun, for instance, then it's easier to ensure that the shadows cast by them will appear correct if their actions are altered.

This will quickly become a political problem. A new wave of journalists are looking to spherical video with a utopian glint. Oh, how this reminds me of the old days. I used to predict that the intensity of spatial video capture would create peace on earth. Amplify empathy. People would really see the awfulness of violence, of war, and would not be able to stomach it. Peace would ensue. We shall see. The more intense a communication technology is, the more intensely it can be used to lie.

Anyway, Graham has devoted the decades of his life since VPL to easing the lives of children confined to hospitals through the use of telepresence. He found an unambiguous way to use technology to make the world better.

AudioSphere

Scott Foster engineered VPL's 3-D sound technology. Since VPL was situated on a marina, it wasn't unusual for engineers to live on boats and occasionally commute by sail, but Scott is the only one I can remember who commuted by small plane. He'd fly in from an airstrip up by Yosemite to a field right next to the marina.

Scott designed custom PC boards to compute three-dimensional sound. People would hear the results through audio headphones incorporated into the EyePhone.

What is three-dimensional sound? It's complicated! You hear the world spatially in part because you have two ears, and the brain can compare what they hear. Sounds arrive at each ear at slightly different times, for instance, and the brain can use that difference to sense the left/right axis direction from whence a sound came. But that's only the first line of perception.

Your brain is also good at deciphering echoes. Not as good as a bat's, but better than you're usually consciously aware. You hear patterns of echoes that convey such news as the shape of the space you're in, what the surfaces in that space are made of, how humid the air is, and where you are in the space.

So the sound subsystem in VR has two obvious chores. It must time arrival of sounds to each ear individually, and it must approximate the echoes that would resound in a real space.

But then there are the pinnae. These are the oddly formed parts of the ear that stick out of the head. Why the irregular, lumped, spiraling shape? The weird sculpture gathers sound coming from in front of you a little better than from other directions, but it also makes sound take on a different timbre—a different shade—according to the direction of origin.

Scott's boards used convolution to simulate the function of the pinnae. Convolution might be thought of as a mathematical metaphor. As an earlier signal was measured to have changed before, so will a new signal be changed—in a similar way—by a convolution algorithm. We use convolution all the time in VR.

In this case, hapless graduate students had to endure being placed in a perfectly echo-free, quiet chamber—an anechoic chamber—with little microphones stuck uncomfortably up in their ears. A speaker blurting test tones was then moved all about. The convolution algorithms then analyzed the sounds that had been recorded inside the ear from particular directions and applied the same transformations to whatever new sounds might be emitted from within a virtual world.

The result was preternaturally good. It turned out that blind VR users could navigate virtual worlds by sound better than in physical environments. Simulated spatial sound could be crisper than reality.

These days, chips are so good and cheap that 3-D sound is just assumed in VR systems. And yet in a lot of the newer systems, the 3-D sound doesn't seem quite calibrated correctly. We get sloppy when things aren't expensive.

No introduction to spatial sound is complete without mentioning the dramatic history of scams associated with it. There's an easy way to get noninteractive spatial sound, and it's been around since the earliest days of audio recording. You put two microphones in a dummy head about where the eardrums would be in a real head. The microphones then gather sounds that have made their way through the dummy head's pinnae.

There are a few classic dummy head recording demos that are rediscovered every decade. The most common one is a haircut. I can't really remember what a real haircut feels like or sounds like, but people who do tell me that the sensation of hearing scissors snipping around your head (from a recording of a barber snipping scissors close to a dummy head) is so vivid that you get goosebumps and chills. (If you're mischievous, you can even get the sounds of the scissors to seem to move into the interior of one's head. All you have to do is tear a hole big enough for scissors to enter in the dummy's head.)

This dramatic demo was used repeatedly over several decades to raise epochal sums from dumb investors who never realized what an easy trick it is.

18. Scene

Demolition

We Veeple spent half our time giving demos and half our time trying to avoid giving demos. The machines were so expensive that it was impossible to devote a setup strictly to entertaining visitors, and yet astonishing pressures were applied to show visitors the prize. A VPL demo was one of the world's rarefied experiences.

Members of the board or investors would demand demos for unexplained visitors in fancy suits; part of the hidden barter of access in the Valley. There were a few times when we were legally obliged to give demos; they were valued enough that they figured into contracts and settlements of disputes. Big customers would demand demos for their employees or customers passing through town.

A typical afternoon demo session at VPL: Terry Gilliam, Monty Python member and film director, has just come to a resting spot within one of my spooky, surreal virtual worlds, called "ritual world," and started talking from inside. He tells me that as I get older, which was inconceivable, I'd have to worry about younger people grabbing the spotlight. "They're surprisingly good," he says.

Then, main receptionist opens the door to the demo room and yells, in a profound Scottish accent, "Go ahead with the Leonard Bernstein demo—the Dalai Lama is stuck in traffic."

Some of the geometrical designs in one of Jaron's early virtual worlds.

Demos were unfair. Certain politicians and celebrities got them more easily than other worthy people, though there was also an informal demo scene late at night. Various counterculture creatures flowed through in the wee hours; snuck into the lab to experiment with combining VR with weird sex or drugs. I'm sure I never knew the half of it.

Demos that by all rights should have been crazy turned out to further the cause of sanity. There was the visit of the faux band Spinal Tap, their wigs carefully suspended in golden buckets (should there be a photo-op). They collided in the lobby with Senator Al Gore, who was on the way in, and Peter Gabriel, who was on the way out. Oddly, the batch of us had a conversation about technology that was unusually well informed and creative. We talked about things like networks and musicians' rights, years before music livelihoods would be decimated by the Internet.

I ended up spending a lot of time with Gore. He pushed for a unification of the various nascent digital networks, and had a breakthrough in 1991. At long last there was funding and momentum to bring all the previously disjointed networks together into one Internet.

It was fun to support Gore. Along with other figures from the virtual reality world, like Fred Brooks, I went to testify in the Senate. It was the first time I wore a suit and tie in years, maybe since my bar mitzvah, and the last time I ever did, as even Senator Gore thought the costume looked ridiculous on me. He got the Senate to adopt a statement of thanks for my sincere attempt at proper attire.

It's sad that Gore was later ridiculed for having supposedly claimed that he "invented the Internet." On the one hand, he didn't say it, but on the other hand, he sort of did invent it, since the lack of an Internet wasn't a technical problem but a political one. The way Gore was smeared turned out to be a prototype for the design of misinformation on the Internet he enabled.

One more story about Gore. Once, right after he had become the vice president, I was visiting him at the Old Executive Office Building in D.C., and these words came out of my mouth: "With the Internet working, everyone will be able to get direct access to the scientific world. Denialism of global warming will no longer be possible." Yep, I said it, and I *so* knew better, especially by then. It just goes to show how easy it is to be seduced by the pop mythologies of one's time.

Who were my favorite demo recipients? I don't really know how to pick favorites. I guess my fondest memories are of kids in wheelchairs who came to experience flying.

Leon Theremin! He had invented the VR-like musical instruments I used to build when I was a kid. He did it back in Moscow, early in the twentieth century. Leon later became a glamorous entrepreneur in America and married a glamorous ballerina; then in the Cold War years he was kidnapped by the Soviets and forced to build spying devices for the state. (He deviously disabled such devices. A recording bug would emit a warning beep, for instance.) No one had heard a thing about his fate for years when the Stanford computer music lab located him and brought him, by then in his nineties, to America for a visit. Leon vibrated so intensely in his glee while in VR that I worried for him, even for the people standing near him.

Demos initiated lifelong friendships. Yoko Ono brought her teenage son, Sean, over for a demo; we've stayed friends after all these years.

The Art of the Demo

VPL had the most powerful graphics computers of the day, but even so we could only offer origami worlds in the early years. It finally got better as the 1980s came to a close. Today's VR demos have a lot more detail, approaching a cinematic visual aesthetic. Even so, the core of the VR experience is the interactivity, and we could deliver that, even back when the visual side of the medium was nascent.

In order to give a great VR demo you have to develop a patter and timing. There was usually a main guide and perhaps a helper, who made sure no one tripped on cables or wandered out of the safe area. The guide subtly tugged visitors through a sequence of virtual world experiences in such a way that they'd be as blown away as possible.

If you're a good VR demo-giver, you chaperone people through virtual worlds in a way that makes it seem that the visitor is completely in charge. Just as a visitor might be flying away from the spot where you've prepared a surprise, like a trapdoor to fall through or a little flower that will suddenly expand into an intricate sculpture when you touch it, you might happen to bump into a gloved hand, apologize, but still manage to knock them back on course. Then they get the payoff.

At the start of the demo, you have to use mime, jokes, and any other trick that might work to get your guest to learn a few rules and skills as quickly as possible. In a lot of the VPL worlds, you'd point to fly forward, point with two fingers to fly backward, and so on.

We discovered clichés that are still staples of VR demos today. A rendering of the physical room the demo is in can be used to marvelous effect, because you can screw around with it. Have the walls suddenly melt inward.

We always planned a moment when the guest passed his head into a virtual object that was startling to behold from the inside. For instance, the person might lean into and through the permeable head of a chromium dragon, suddenly to be enmeshed in gears and cables whirring away right at one's face, right into the skin. People yelped and shivered.

I still see that trick today, as well as the one where you get huge and your partner gets tiny and walks on your hand, or the one where there's a pit in the floor that you're afraid to step on, even though you know it's not real. (Mel Slater invented that pit.) Jeremy Bailenson once commented that the pit had become the first iconic trope of VR, like the Lumière brothers' oncoming train.

After a few years of practice, we learned to show our visitors the ropes so quickly, it felt as if they'd always known. We had no shame. There were usually unannounced tummlers in the room cheering visitors on for the first few minutes as they made virtual toddler steps.

As hard as it might be for present-day VR-heads to believe, given how crude they look in old videos, our demos were utterly absorbing.

VR is finally cheap and commonplace enough that the culture of the VR demo is fading. People just experience VR worlds at home or wherever they are without any guidance from the creators. We should make a point of enjoying the culture of demos where it still exists in special circumstances; in classes, at trade fairs, or when teams are pitching designs.

The best trick for enhancing a VR demo is to sneak genuine flowers into the vicinity while visitors are inside the VR experience. They'll come out and experience a flower as if it was the first one they've ever seen. The best magic of VR happens in the moments right after the demo ends.

The Art of the World

Since VPL's customers were mostly industrial or academic, most of our virtual worlds were useful instead of fantastical. You'd learn to perform surgery on a knee, repair a jet engine, design a kitchen, and so on, all with rudimentary graphics, yet demonstrating an unprecedented path to understanding and skill acquisition.

But what many of us really wanted was strangeness. So we Veeple rose to the occasion and made strange worlds for pure joy, even though it was hard to find the time.

I liked giving animal essences to conventionally inanimate objects. Clouds and desks would get tails that whipped about playfully. Details of my worlds reflected the user in subtle and unexpected ways, so as to blur the line between avatar and environment. The chandelier might sway according to how you shrugged your shoulders, just at the edge of perception, so that you might not consciously notice. My colors were always slowly changing. You might find yourself getting gradually larger or smaller, once again at the edge of notice. I refused to incorporate any rectilinear element into my virtual worlds, recalling the Earth Station.

There was one world we showed at SIGGRAPH in which Ann and Young's little daughter turned into a teapot avatar—the famous computer graphics teapot*—in front of the big film show audience. Yes, she sang "I'm a little teapot." We were shameless.

Ann was the absolute mistress of whimsy. She made a funny and spectacular Alice in Wonderland world, for instance. You could run into the

---

* For years everyone in computer graphics used the same teapot model to demonstrate rendering techniques. You can even see one in Pixar's original *Toy Story*.

White Rabbit's mouth. It was a loving, Klein Bottle* elaboration of John Tenniel's original illustrations.

You can't aim a selfie stick in San Francisco lately without catching people in the background who turn out to have VR design startups. They'll sometimes ask for my thoughts on virtual world design, and as you must have guessed by now, I tend not to remain silent.

Advice for VR Designers and Artists

a) Your most important canvas is not the virtual world, but the user's sensorimotor loop. Stretch it, shrink it, twist it, interlace it with loops from other people.

b) Emphasize biological motion over rigid UI (user interface) elements that throw away most of what the body does. The worst offender is a button. Avoid buttons. Use continuous controls.

c) We already have VR clichés, so it's already worth avoiding them. Enough trapdoors, objects flying at your face, things that change when you look away and then back. Or, embrace clichés but find a way to make them serve a larger purpose.

d) Test your world with diverse people. Better yet, add diverse people to your team. Cultural background, age, sex, and cognitive style have a bigger impact on how people take to VR than to other media. Make sure you understand how your design fits into the broader landscape of human cognitive style, because that's the only theater in which it means a thing.

e) Corollary to the previous rule: The science of VR is young, so question received wisdom about who VR is for. If you're told VR works better for men than women, you should wonder if that's because the virtual worlds tested were designed by men.

f) Your overriding narrative arc is not within a virtual world, but in the real world; it's the one in which a person starts to engage with your design, engages, and then ceases. This might be when he or she puts on a headset, does whatever is to be done, and then takes off the headset. Think about that overall experience. What is the person expecting before going in? What's it like to come out?

---

* A Klein Bottle is a beloved and weird geometric shape; a bottle that is inside itself.

g) Experiment with resisting whatever is easiest to achieve with your development tools.

h) Think about the other people who are around but not in the virtual world. Are they part of the experience? Are they watching what the immersed people see on an old-fashioned screen? Is there any common cause between the people who are inside and those outside?

i) Fight against impulses you internalized in film school. VR is not cinema. For just one example, the watcher becomes invisible in a movie, but not in VR. The navigable virtual world is less important than the body of the user. What does she see when she looks at her hand? In a mirror? If the answers are modular—not central to the story—then you aren't yet designing for VR.

j) Fight against impulses you internalized from gaming. For just one example, a game that is thrilling on a traditional screen can come off as kind of isolating and dull in a VR headset. One reason is that a person is bigger than a game on a screen, but smaller than a game that surrounds, so the calculus of personal status inverts in the usual games about chasing, shooting, and being chased and shot.

k) Users must be able to leave a mark, dent the universe. Otherwise they won't really be there in full, meaning that you won't have succeeded in designing a virtual world.

l) Don't assume everything must be algorithmic and automatic. What if there's a great niche for a live performer in your world, maybe appearing over the 'Net, maybe even to get paid?

m) Do think about hazards and safety. If you're working with a tethered headset, for instance, think about whether motions contribute to tripping hazards. This applies even if people remain seated. Don't let them spin in the same direction endlessly, for instance. Be honest with yourself and your users when a design presents an unusual likelihood of simulator sickness; if that's so, let people know they shouldn't drive for a while after.

n) Do worry about power dynamics and potentials for confusion or abuse. But not at the expense of daring thoughts about how the future can be better. Be tactically pessimistic and strategically optimistic.

o) Don't necessarily agree with me or anyone else. Think for yourself.

Forty-third VR Definition: A new art form that must escape the clutches of gaming, cinema, traditional software, New Economy power structures, and maybe even the ideas of its pioneers.

Flags Planted

It's often said that I coined the term "virtual reality." It depends on how you think about the boundaries between context, languages, and history. There's a wonderful argument that I did not.

Before World War II, the radical dramatist Antonin Artaud used the French phrase *réalité virtuelle* in his discussions of a "theater of cruelty." This wasn't a nasty notion; what Artaud meant was a nonverbal form of theater that was intense enough to rouse depths of human experience and understanding beyond the reach of conventional language.

I started using the phrase before I knew about Artaud, but I couldn't be happier about a connection across generations. Present-day VR-heads would be startled to read either Susanne Langer (who came up with "virtual world" in the 1950s) or Artaud.

There are other disputes about the origins of VR vocabulary. I distinctly remember the science fiction writer Neal Stephenson coining the term "avatar"—not as a word, obviously, since it has ancient origins in Hinduism, but as the term for your body in VR. And yet, apparently, there are competing claims.

"Virtual reality" isn't the only term for what it roughly describes, and it's hard to believe how intensely people fought over words back in the 1980s. Terms took on tribal significance.

Partisans would try to get a session at a conference to be named "virtual environments" instead of "virtual reality" or vice versa. "Synthetic reality" and "artificial presence" were also team banners, though I can't remember who was on whose team anymore. It's hard to believe in retrospect that anyone cared about such things.

Myron Krueger, another pioneer in our field, preferred "artificial reality." Back in the 1970s he would render the outlines of people's bodies into a TV screen in real time so they could interact with artificial objects. It was remarkable early work, foreseeing familiar modes of interaction such as what we do with Kinect sensors today.

"Virtual environments" was a term associated with the "big science" places like NASA, so a lot of the formal literature from the period uses that term. It might have been coined by NASA's Scott Fisher.

"Telepresence" used to mean being connected with a robot in such a way that you felt as though you were the robot, or at least that you were in the robot's location. The community that studied telepresence had started way back in the analog era, well before Ivan Sutherland, or even Alan Turing. Lately it has a broader usage, including Skype-like interactions in VR or mixed reality.

"Tele-existence" was coined by the wonderful pioneering Japanese VR researcher Susumu Tachi to include both telepresence and VR.

I wish I could remember the precise moment when I started using the term "virtual reality." It was in the 1970s, before I came to Silicon Valley, and it served as both my North Star and my fledgling calling card.

VR was the term I liked for first-person presence in a virtual world, but most especially when there were other people in there with you. In a technological setting, "reality" could serve as the social version of Ivan Sutherland's "world."

The hippie culture of the 1970s was obsessed with the idea of "consensus reality." I've always been infuriated by sloppy New Age philosophy, probably because when my thinking fails, that's a cliff I risk falling off of. It was common in the 1970s to hear assertions about how, if all people everywhere could only believe something all at once, anything could be changed. The sky could become purple, cows could fly. Reality was only a collective dream project, and tragedy was the fault of bad dreamers.

It seemed to me that underplaying the reality of reality only ended up obscuring the more useful aspect of that way of thinking: If everyone's thoughts could change, maybe the world could become kinder and smarter. And yet even in that case it is not necessarily easy to know what people should think or dream. There is an unavoidable problem-solving aspect to making the world better.

True believers around Silicon Valley used to demand that we all dream of socialism, but then the demands were for libertarianism, and lately they've been for the supposed supremacy of artificial intelligence. No one is ready to accept that the perfect dream hasn't been articulated and might never be.

At any rate, the word "reality" had more than a touch of a utopian charge in the 1970s, and I liked that feeling, if not necessarily all the cultural baggage.

So far as I know, I also coined "mixed reality."* However, in that case one of our biggest customers, Boeing, had an engineer who became enamored of "augmented reality" instead, so of course we were happy to use that term. I still like "mixed" better. Maybe "stirred"?

Lately, "augmented" means you see the world annotated, while "mixed" means you see extra stuff added to the world that can be treated as if it were real.

There was a branding value in "virtual reality," since it was originally associated with VPL Research. Not everyone at VPL loved it. Chuck, our alpha hacker, thought it sounded too much like "RV," short for recreational vehicle. "It sounds like we want old people to be put away in simulations so we don't have to deal with them." Hope he turns out to be wrong about that.

Anyway, another definition:

Forty-fourth VR Definition: The term you might have used in the 1980s if you were partial to those weirdos at VPL Research.

Moniker on the Loose

Long after the events chronicled in this book, the fall 1999 edition of the *Whole Earth Review* included my piece about the widespread use of the term "virtual reality" at that time to mean many things. Here are some lightly edited excerpts:

> Decades ago, I called a type of computer-user interface technology "virtual reality." Two qualities, social and somatic, together created something

---

* An example of my 1980s usage of the term is found in *Virtual Reality: An Interview with Jaron Lanier.* Kevin Kelly, Adam Heilbrun, and Barbara Stacks. *Whole Earth Review*, Fall 1999 n64 p108(12).

quite different from a solitary virtual world. VR functioned as the inter-stices or connection between people; a role that had been previously taken only by the physical world. The term "reality" was appropriate.

A "world" results when a mind has faith in the persistence of what it perceives. A "reality" results when a mind has faith that other minds share enough of the same world to establish communication and empathy. Then add the somatic angle: A mind can occupy a world, but a body lives in a reality, and with VPL's somatic interfaces like gloves and body suits, we were designing for the body as well as the mind.

Pop Culture Fantasia

We now leave the actual technology behind and follow the adventures of a metaphor as it skips out of the lab into the big wide world. Virtual reality's potency as a metaphor is so great that it is almost impossible to track.

Here is an incomplete survey of the current usage, as of summer 1999:

A delinquent disassociation from the truth: In the last presidential election, each of the four national candidates at one time or another accused his opponents of "living in virtual reality." This was when they were being nice; at other times, more aggressive language was in play.* "Virtual reality" meant the failing of a mind that might be well-intentioned and clever, a delusion rather than a manipulation.

A protean, all-encompassing triumph of creativity: The cover of a Frank Sinatra CD brags that "Frank creates a virtual reality when he sings." The term has appeared endlessly in blurbs for novels, movies, and recordings.

Pervasive alienation: Distance from natural reality brought about by technological civilization. Rather than the mere Marxist alienation from one's work, a person is alienated from all of natural life by the devilish confusion of mass media and other pervasive technologies. I

---

* The four candidates were Bill Clinton, Al Gore, Bob Dole, and Jack Kemp, and nothing any of them said would be considered aggressive by today's standards.

was given a Gen-X faux-fifties refrigerator magnet with a Norman Rockwell family overlaid by the words *VIRTUAL REALITY* in a sinister font. In this usage, virtual reality is essentially treated as an ultimate form of television by people who hate and fear television.

An ecstasy or epiphany brought about by technology: The first cover story about the technology in the *Wall Street Journal* referred to it, incredibly, as "electronic LSD."

A transcendent perspective brought about by technology: Hollywood scripts have frequently used virtual reality as a device to give a character, and the audience, privileged knowledge. The idea is that he with the goggles sees farther. In the early days (*The Lawnmower Man*), the knowledge was often used to either rule the world or solve a crime, while in more recent incarnations (*The Matrix*), the hero uses [the ability to escape from] virtual reality to become a Buddha-like or Christ-like figure, wiser than the common mortal.

The Ambiguity at the Core of the Metaphor

Now why would a metaphor about a user-interface technology take on such grand pop overtones? I think the reason is that "virtual reality" evokes unresolved mysteries about the status of computers and all things digital.

Computer scientists might think of the whole world as a big computer or a collection of algorithm-organisms, which might be trees or humans. They have entertained the public with the question: Is there ultimately a difference between reality and a very good computer?

This explains both sides of virtual reality as a pop metaphor. Virtual reality is transcendent, because if reality is digital, it is programmable. Everything becomes possible. You can enjoy a universe as varied as dreams and still share it with other people who are plugged into the equipment, instead of being trapped in your own head. To all those connected, a tree can suddenly transform into a sparkling waterfall.

On the other hand, if reality is digital, everything is the same as everything else. Claustrophobia quickly sets in: A bit is a bit. As you watch the tree change into a waterfall, you realize there was nothing essential about the bits being a tree or a waterfall, or you being you, for that matter.

The Party

There was a party scene, more than that; a whole cultural infrastructure connected with virtual reality. I find it a little painful to describe because it was awkward; embarrassing. It was like space travel dreamers confining themselves to a fake spaceship in a lonesome desert because a real spaceship will not be available in their lifetime.

A typical happening would not involve actual VR experiences, for the equipment was rare and expensive. Instead, guru candidates would talk about VR. There would be speaker after speaker, plus VR-themed bands, weird party decorations, strange locations, all evoking what VR might be like someday.

That old world of speculative VR obsession, the psychedelic tech party circuit, evolved into today's Burning Man festival, or at least Burning Man as it is at night, when you no longer see the mountains but only the blinking lights of human invention. A simulation of what it might be like to be able to improvise reality fully; a simulation of a simulation.

Remembering the era of VR-themed parties evokes feelings of guilt and anger in me, even today. Anger because there was an excess of guru wannabes who aggressively pecked at me because they wanted the prominence. It was like one of those obscure art world or academic microcosms filled with jealousy and backstabbing, even though the stakes were preposterously low. I found camaraderie with other people who were actually building VR tech, but the talkers included a lot of useless self-promoters. Charlatans used the scene to hawk bogus medicine and other sideshow scams.

The guilt arose because there were lovely young folks who cared about the scene a great deal and found a way of life in it—and plenty of those were not shy about accusing me of betrayal when I eventually stopped showing up.

I came under severe social pressure to make VR demos available at VR parties. Once in a while, at major events, VPL would allow little cabals of partyers to trickle in for demos. The big gathering would be in a decaying unused factory, an abandoned ferry, or some other eerie Bay Area party environment, and a few people at a time would come in a secret van from there to VPL's offices by the bay, all through the night.

A stable of speakers and bands became established. (My favorite of the bands was called D'Cuckoo. Linda Jacobson was one of my favorite GNFs and VR pundits.) The VR party universe overlapped with the psychedelic one and the Grateful Dead one; it drew from the fractal, endless catalog of utopian crews and cults around the bay.

In a sprawling, haunted-beautiful nineteenth-century wooden mansion above a gurgling spring in the Berkeley hills there lived a circle of roommates who published esoteric psychedelic magazines. They adapted to the VR party aesthetic by concocting a tech magazine with a psychedelic style, called *Mondo 2000*. (The numeral 2000 conveyed the impossibly distant, undoubtedly transcendent and terrifying future.)

Mondo was the prototype for much of what has become the familiar, intoxicated-by-neoteny style of the Valley ever since. Brightly colored psychedelic goofiness. Nonsense rhyme names for whatever is new in the world. An over-the-top toddler's fantasy of superpowers. *Wired* sure looked like a rip-off of *Mondo* when it first came out, but it was cool; the early *Wired* people were part of the circle.*

Ricochet la Femme

Guess who was living in the *Mondo* house? The woman I met back in Cambridge, who I told you I would marry.

Only a few years after our first meeting at MIT, I had become well known. I was the answer to *Jeopardy* questions on TV and appeared on magazine covers. Suddenly in the game.

She whispered, "You will bring about a revolution in the story of mankind. You will change communication, love, and art. I will be at your side."

We got married. Probably the worst mistake of my life.

I have special trouble reconstructing dialogue with her because I just wince too much. What was I doing?

There's a stratum of self-important men and women who make each other feel like the importance is a real thing. When I became famous, in the late 1980s in Silicon Valley, I saw and felt what the maelstrom of sex and power is like; a hidden world of titans jousting, whales and giant squid.

---

* *Wired* renewed the morphology of the early literature of computation: One half was nerdy systems thinking, with a utopian sensibility and a realization that nerds are running the world now, while the other half, which I liked better, was psychedelic revelry from a personal perspective. I was on the masthead as a contributing editor in the early years.

Young women would spend hours prettying themselves up to make powerful men feel mythic, generally in exchange for crumbs.

Later on I got to know a few women who played this game, this time as a friend instead of a combatant. They were often accomplished, perfectly capable of taking care of themselves, but even so, they sometimes found the gravity of ancient clichés to be inescapable. One I knew clung to a certain Donald Trump for a while. "He makes me feel safe, like he will protect me." He treated her terribly and dumped her crudely.

But this is about my life, and the truth is that I had moments later in the 1980s when I got pretty full of myself. I was asking for it.

The feeling of combining romance with self-importance is so powerful that it realigns reality, even in the perception of people around you. Like Steve Jobs's famous "reality distortion field."

This wasn't lust, exactly, but something more powerful; a deep human business, ancient, like discovering a hidden sexual organ that communicates with the great men of history and brings you into their immortal communion. Inner demons of vanity congeal into a seductive monster that envelops you and intones, "The great scientists and conquerors, the ones we remember, you will join their ranks."

It's so dumb that I can hardly stand to talk about it, but I hope that pointing to a giant sinkhole might just break the spell for someone else. I wonder what could have broken the spell for me at the time.

I lived with her in the *Mondo* house for a while. She got into a gargantuan tiff with one of the prima *Mondo* editors, Queen Mu, who took up most of the refrigerator with samples of what she said was tarantula venom. I can't remember what that substance was supposed to do to you. Said my wife, "If women ran the world, there would be a lot fewer wars, but many more poisonings."

We moved out to a nearby faux Greek Temple, garlanded, resplendent, that had been built by Isadora Duncan's circle.* Those days were like living in a Maxfield Parrish† painting. A pageant of variegated exotica. Then we lived briefly in a dramatic, expensive house overlooking San Francisco; a movie scene, a shrine to significance.

She wanted to get married, but she talked about it as though it was a

---

* Isadora Duncan was an early luminary of modern dance and a notorious free spirit who came from the San Francisco area.

† An influential American painter known for the dreamiest scenes ever rendered.

prize, a touchdown, a royal flush. I don't think of her in retrospect as an antagonist, but more as a victim who fell into a deep trough carved out by trauma and tradition. A comically exaggerated gold digger personality, an archetype from central casting, emerged in her; a mirror image, I suppose, of the stupid vanity monster that emerged in me. Her demons dragged my demons to the courthouse one day, and while it might have looked like happiness, I actually wept in shame and fury throughout the ceremony. Both she and I had lost battles with horrid, inherited pseudodesires, not really our own.

Was the marriage entirely bullshit? Not quite.

Limerence can exist apart from lust. It can be made of narcissism, ambition, phantom limbs of childhood unlived. The texture of life became so intense; saturated colors, fragrances so sweet as to knock you out. I remember those feelings like a theory now, a structure, a placeholder for luminous curiosities that will never return.

What was most remarkable about my weird little first marriage was that by going through the experience of self-immolating infatuation without really being attracted to the person in question, I felt a pure form. To put it in nerd terms, I felt the exposed power of romance as if it were computation, the genetic engineering that formed us and creates the future of life. Limerence might be a transient vapor, but there is something there: an entanglement with life writ large, the billions of years of it, the great structure in which you are a tiny bud, or the makings of mulch for the next bud.

But every little bud really does steer the billion-year blossoming a little. Romance might make us into helpless fools, but we are also creating; we are artists of the universe. I felt that. Maybe the whole horrible experience was worth it.

Dark Lineage

There was a fresh literary scene related to VR in the 1980s called cyberpunk. It was, in my view, a continuation of E. M. Forster's "The Machine Stops." Dark stuff, usually; cautionary tales.

Characters typically manipulated and deceived one another, or wallowed in existential malaise. Vernor Vinge wrote a novel called *True Names* and a little later came *Neuromancer* by William Gibson.

I adored *Neuromancer*, but I had the ridiculous idea that I was called upon to brighten the cyberpunk movement. It'll be a murky mess, trying to reconstruct old dialogue between me and Bill, but here goes:

"It'll attract people even though it ought to repel them," I'd say. Bill was game to talk about it. He still sounded like he was from Tennessee. Canada would eventually smooth out his accent a bit.

"It's not like you calculate a book, Jaron. It comes out. When I was a kid, I was blown away by *Naked Lunch*, and I try to imagine a kid reading *Neuromancer* and being blown away."

"*Neuromancer* is definitely blowing away young versions of you, no question. But couldn't you try to come up with a more positive future, something to aspire to, since what you're doing is making all this stuff seductive, but it's a bummer?"

"I could try, Jaron, but this is what comes out."

"I just worry that darkness never seems to serve as a warning when it comes to computer stuff in science fiction. The dark stuff just comes off as cool and people want it."

"My job isn't to fix humanity. You can give it a try; you're actually building things."

"Oh, thanks."

"If I had it to do over again, I'd probably try; to have a VR startup instead of writing novels."

"You're welcome to come and work."

"Um . . ."

At the time I had no idea how hard it was to write a passable book, much less a good one. I wish I'd let Bill alone.

Other great cyberpunk writers appeared. Bruce Sterling came across like a young Hemingway with a Texas drawl. Neil Stephenson was our Apollonian scholar.

If you pay attention, you'll find cameos of me in early cyberpunk novels. My head might float by.

Flattering Mirror

Fiction about VR has mostly been quite dark ever since cyberpunk. The *Matrix* movies; *Inception*. Meanwhile, norms for tech journalism became hell-bent on positivity.

VR engaged a new generation of journalists, like Steven Levy, Howard Rheingold, Luc Sante, and *Mondo 2000*'s Ken Goffman, aka R. U. Sirius. I'll highlight two figures who were particularly influential as well as dear to me: Kevin Kelly and John Perry Barlow.

Kevin is a fine example of a trusted friend with whom I disagree completely. When I met him, he was editing and writing in publications connected to Stewart Brand's world, post–*Whole Earth Catalog*; he later became the first editor in chief of *Wired*.

Kevin thinks that objects we perceive to exist in software really exist. I do not. He believes in AI, and that a noosphere not only exists, but might have gained a kind of self-determination now that computers are networked. I do not. Kevin thinks technology is a superbeing that wants things. He perceives grace in that superbeing. I was delighted to provide a blurb for his book *What Technology Wants*, stating that it was the best presentation of a philosophy I didn't share.

Kevin remembers that we all came upon our ideas just now, let's say three minutes ago. We shouldn't treat our ideas about computation as hallowed. He has a sense of humor and an open mind.

John Perry Barlow claims to have a perfect memory of meeting me at a hacker retreat, but I can prove I wasn't there. It's weird, because he's supposed to be the one with crystal-clear memory of everything, and I'm supposed to be the one living in a fog.

Barlow and I got close fast; we have a lot in common. He had been a rancher in Wyoming and found fancy city life to be as much of a put-on as I did. We loved reading and writing, which was more of a novelty than it should have been in the tech scene. Barlow worked in the music business, so we had friends in common from that world.

He was a lyricist for the Grateful Dead, which was more than a band in those days. It was a lifestyle for its fans. So Barlow was revered and lived an elevated life.

We had different approaches to socializing. Barlow lived as if he was always on camera; always holding court in one way or another, always careful to make each utterance memorable. A ladies' man, always strategizing.*

---

* This is not my judgment, but how Barlow describes himself: http://www.nerve.com/video/shame less.

I refused to participate in Barlow's scenes. I'd only see him one-on-one or with another person or two who were true friends and not hangers-on. With those ground rules in place, Barlow and I grew close, and I came to love the guy.

Barlow initially wrote about VR as a gonzo journalist. It was fun. Later, he entered into a magnifying resonance with ideologues for one of the putative digital utopias.

That development was difficult for me.

Virtual reality had been dubbed cyberspace in *Neuromancer*; remember, the rule was that everyone had to come up with their own term for it.

Barlow took Bill Gibson's term and recast it as the name for what he perceived to be the reality of bits.

Later, in the mid-nineties, Barlow would pen a declaration of independence for cyberspace. It was to be a new Wild West, but infinite, and forever beyond the reach of governments, a libertarian paradise.

I thought Barlow's redefining of cyberspace was a mistake, but it wasn't worth arguing about. There was plenty of room for all our ideas. I didn't want to reenact the "splitism" about ideas that made Marxists ridiculous. But Barlow was an organizer. He eventually put me in a position where I had to choose.

19. How We Settled into a Seed for the Future

Virtual Rights, but Not Virtual Economic Rights

In 1990, I was invited to a lunch at a Mexican restaurant in San Francisco's Mission District to consider cofounding a new organization to fight for cyber rights. Chuck, VPL's prime hacker, and I went up and met Mitch Kapor, John Gilmore, and Barlow. The three of them eventually moved forward, founding the Electronic Frontier Foundation.

But I held back. (Chuck was too busy coding to pay any of us much mind.)

I didn't say why at the time; wasn't ready to state my doubts to these sweet friends. I support most of the cases the EFF takes on, but not the underlying philosophy.

The EFF was to support "privacy," such as the right to use secure encryption, but not the ability to prevent others from copying one's information if it can be gotten at all.

The early example was music. The new utopia was to be one in which music that had previously only been legally copied with the payment of a royalty would now be copied "for free."

I felt that you can't have privacy without also forging a new form of private property in the information space. That's what private property is *for*.

There has to be space around a person for a person to be a person. If everything you share at all is suddenly commoditized by whoever has the biggest, baddest network computer, then you're doomed to be a spied-upon information serf. The promotion of abstract rights without economic rights would be nothing but a cruel trick we'd play on those who would be left behind.

I argued that making music "free" would just result in *no one* being able to make a living when automation would eventually advance. If the only value left is information (once robots come to be perceived as doing all the work) and information is to be "free," then ordinary people will become valueless, from an economic point of view.

Of course, that perception of robots doing work would be a lie, because robots don't actually do anything on their own, or even exist apart from people. My sensibility about robots and artificial intelligence is so important to my story that I will convey it in two different ways. Later in this chapter I'll recall the way I used to argue about the topic, while some of my current thinking appears in appendix 3.

The key point is that digital idealism took a turn for the absurd around 1990. We started to organize our digital systems around bits instead of people, who were the only agents that made bits mean a thing.

The Easy Road to World Domination

An online framework called the World Wide Web appeared in the early 1990s and quickly gained traction. When a design starts to win in a digital network, it tends to keep on winning. Even when the WWW was only a tiny nascent undertaking, it soon became clear that it would swamp us all.

Part of the reason was a lowering of standards, at least from my point of view. The WWW introduced one little change in network design that made it the perfect vehicle for the "cyberspace" way of thinking.

Earlier designs for networked information had required that records of provenance be maintained. Any information accessed online could be traced to its origin. If there was a link between one thing and another thing on a network, the link went both ways. For example, if one person could download a file, the other person, from whom the file was downloaded, could be notified of who was doing the downloading.* Therefore, everything

---

* At the time, one reason many hackers supported one-way links was a concern for privacy.

that was downloaded was contextualized, artists could be paid, scammers could be identified, and so on.

Previous designs were centered on people, not data. There was never a need to copy information because one could always go back to the source, associated with a person. Indeed, copying was considered a crime against efficiency.

Tim Berners-Lee chose to offer a different approach with the World Wide Web, one that was much easier to adopt in the short term, though we've paid dearly in the long term. To get started, one simply linked to online information, and the link went in only one direction. No one could tell if information had been copied. Artists wouldn't be paid. Context would be lost. Scammers could hide.

But Tim's approach was profoundly easy to get into. With almost no overhead or maintenance or responsibility to anyone else, anyone could post a website, made instantly substantial by assemblages of material from yet other people.

The World Wide Web went viral, to use the contemporary term. That's not the way we talked at the time. Terms like "viral" and "disruptive" still sounded negative and destructive. We hadn't yet hypnotized ourselves into Möbius-Orwellian tech talk. Now we describe what we are doing accurately, but we pretend we're being ironic, so that we can feel better about ourselves. Shall we call it "Notwellian?"

Wouldn't it be bad, they asked, if someone could trace who got what information? Wouldn't that lead to a surveillance society? The counterargument I made was that if information was made worthless through anonymous copying, then most people would be disadvantaged as technology became more information driven, since they wouldn't receive credit for their contributions. That would cause all power and wealth to concentrate with whoever ended up with the most powerful online computers, and those overlords would still be able to trace everyone everywhere, because they'd end up controlling the network.

I hated to have to think that way back then, and I hate that it now looks like I was right.

There's a subtle way that one-way links undermine capitalism, in particular. Think back to the slimy/attractive yuppie in Santa Cruz who made money by denying other people access to information. I like markets and capitalism; they offer a way to avoid the serious problem identified by Oscar Wilde (too many meetings) without yielding to dictators. But a market works only if each party has information that's distinct from everyone else's. That distinction is part of what gives diverse players diverse opportunities in a market. If certain computers can accrue more information about ordinary people than the ordinary computers those people use, then whoever owns the top computers will start to accrue extreme wealth and power. We see this effect today in companies like Uber. Working-class jobs are made less secure, while the few people who run the overseeing computers become spectacularly rich. The World Wide Web was and remains profoundly antimarket in the big picture even as it created stupendous opportunities for the people with the biggest computers. This stream of ideas is expressed more fully in *Who Owns the Future?*

I remember looking at the first Web pages with people at Xerox PARC, and with Ted Nelson. "Unbelievable that someone would launch a design with only one-way links." That was the universal appraisal; it was cheating. But there was undeniable action there on the nascent Web, more than anywhere else.

We techies collectively acquiesced; we succumbed to the decision to make online networks artificially mysterious by leaving out the reverse links. Maybe we feared that a knowable 'Net would not be commensurate with our capacity for wonder, as it was put long ago, so instead we chose a murky, unknowable 'Net.

By not having two-way links, there was no way to know comprehensively what pointed at what; thus an entirely human-made artifact took on a trackless quality, as if it were a wilderness. The Wild West reborn! But it was only artificially so.

We felt guilty using the Web at first. It must be hard for people who have grown up with the Web to appreciate that feeling.

Much later on, companies like Google and Facebook would make hundreds of billions of dollars for the service of partially mapping what should have been mapped from the start.

This is in no way a criticism of Tim Berners-Lee. I continue to admire and respect him. He didn't have a plan for world domination; only a plan to support physicists at a lab.

Despite the feeling of guilt, the rise of the Web also felt miraculous. I used to wax rhapsodic about it in my lectures. The first time in history that millions of people had cooperated to do something not because of coercion, profit motive, or any influence other than the sense that the project was worthy. Well, actually in retrospect, there was and remains an excess of vanity as motive, but even so; what a remarkable moment to support a touch of optimism about our species! If we could fill the Web out of thin air, maybe we could solve our big problems.

I still feel that sense of miracle, but what made it float was vacuousness. The long-term price we have paid has been too high.

Microgravity

In the early days of the popularization of the Internet, there was a debate about whether to make online digital experiences seem casual and weightless or whether to make them feel serious, with costs and consequences. For instance, early luminaries including Esther Dyson and Marvin Minsky advocated micropostage for email. If people had to pay for email, even if only a tiny fraction of a penny, big-time spammers would be discouraged. Meanwhile email would be appreciated for what it is, a big human project that costs a lot.

The critics of email postage won that debate. They argued that even the slightest amount of postage would disadvantage people who were too poor even to have a bank account, and they had a point. But beyond that, there was a massive desire to create the illusion of weightlessness on the Internet.

In the service of weightlessness, Internet retailers would not pay the same sales taxes as brick-and-mortar ones; cloud companies wouldn't have the same responsibilities to monitor whether they were making money off copyright violations or forgeries. Accountability was recast as a burden or a friction, since it costs money; an affront to weightlessness.

Meanwhile the Internet would be designed as minimally as possible, so that entrepreneurs could experiment. The Internet as a raw resource provided no hook for persistent personal identity, no way to conduct transactions, and no way to know if anyone else was who they claimed to be. All of those necessary functions would have to eventually be fulfilled by private businesses like Facebook.

The result, in the coming decades, was a mad rush to corral users at any cost, even at the cost of caution and quality. A slogan from turn-of-the-century Silicon Valley adapted Bobby McFerrin's famous song "Don't Worry, Be Happy," to exhort "Don't worry, be crappy!"*

We ended up with an uncharted, ad hoc Internet. We made our lives easier during the period described in this book, but the whole world is paying a heavy price many years later.

For one thing, we don't trust the Internet.† Each tech company and service provider lives in its own universe, and the ragged cracks between those universes provide handholds for hackers.

---

* http://guykawasaki.com/the_art_of_inno/

† President Trump advised citizens to not depend on the Internet. "If you have something really important, write it out and have it delivered by courier, the old-fashioned way." Has anything like that ever happened before? Did a president ever discourage citizens from using cars back when the automotive industry was one of the nation's leading lights? (http://www.cnn.com/2016/12/29/politics/donald-trump-computers-internet-email/index.html)

There is nothing intrinsic to computing that makes it sloppy or unserious. The system for online transactions between banks is reliable, for instance, and nobody has ever hacked or leaked the all-important algorithms that run companies like Google and Facebook. But we chose an unserious network.

The Invisible Hand Improves When It Is Made Visible as an Avatar Hand

The appeal of weightlessness arose out of a desire to make digital networks into an instant permanent solution to an eternally unsolvable problem. Finally, it would be possible for people to connect without the tedium and annoyance that always must attend cooperation between genuinely free, distinct individuals. The dream was to launch a form of democracy unburdened by politics. Freedom unburdened by other people's rights. Anarchy without peril. The only way to do it was to make people less real.

It amazed and saddened me that so many digital thinkers internalized their intense, formative experience of virtual reality in a way that contributed to what I see as the great confusion of the Information Age. Barlow was only one of many. He moved directly from a fascination with VR to what I saw as a terrible formulation of the ideal Information Age society.

Perhaps our different views were related to my background in farming versus Barlow's in ranching. Fences were his enemy, but my friend.

The cyberspace way of thinking about bits in a network suggests that it's a place where you float. You don't count on help, and you have no responsibilities; you are free to roam. You get to pick the fruit of the land, the free content and services.

It's the cowboy idea, reflected in the inclusion of "frontier" in EFF's name, and even more so in Barlow's famous "Declaration of Independence for Cyberspace."*

At least it's a scheme that actually benefits hackers; the real Wild West (not the one in the movies) was rarely generous to horsemen or gunfighters. Ultimately, though, the top beneficiaries are those who own the biggest cloud computers, just as the real Wild West was all about who owned the railroads and the mines.

---

* https://www.eff.org/cyberspace-independence

Barlow was a moderate! He caught a lot of grief from hackers who wanted an even more extreme vision of hacker supremacy.

Before the World Wide Web existed, there was a ubiquitous bulletin-board-like service called Usenet. It had been around since 1980, *way* before the Internet,* so it was already a somewhat stale institution by around '87. That was when it was reorganized by a few people, including John Gilmore, who would become a cofounder of the EFF, to support a chaotic explosion of user-created topics. The anarchic new universe was called the *alt. hierarchy.*†

Surprise, there was a lot of porn. But this other thing started to happen. Conversation threads in the alt. universe started to get extreme. It's not just that the most reprehensible people in existence—pedophiles, for instance—had found a forum.

Reasonable people started to change through online experience, and for the worse. I knew a few hackers who had previously dwelled only on the edge of nuttiness; online they plunged into a pattern of mutually reinforced crankiness, conspiracy theory amplification, and bullying of anyone who didn't agree. What was only a fringe phenomenon at the time turned out to have world-changing consequences decades later.

To be clear, most of alt. was great. I used to talk about weird musical instruments on it. But the nasty edge was loud and unavoidable. Spam was born. The troll persona appeared and flourished.

A new medium was bringing out the worst in a tiny minority of people, but that minority was in your face. We suddenly had a new global network of bridges between people, but even if trolls lurk under only the occasional bridge, that changes the way you cross every bridge.

Political discussion groups coalesced and drove themselves into ever more irritable inflammations. It didn't matter so much whether the ideas came from the left or the right, as long as the packaging included extensive, snide faux-technical expositions and outsiders were vilified. Astonishing levels of antagonism built up toward designated losers, usually based on sex or race.

---

* Networking existed before the Internet, but it was fragmented. The Internet was the political advent of network interoperability.

† It was called a hierarchy because of the tree of subdivisions into topics. For instance, there was one leaf on the tree called alt.arts.poetry.comments and another called alt.tv.simpsons. About twenty thousand groups like these are thought to still be active in 2017.

(The story of how our scale model of a degraded society exploded into mainstream politics and society is told in appendix 3.)

I often replay those years in my head. There was one argument, the one about censorship, that drowned out every other notion about how to improve the online world. It was framed in stark all-or-nothing terms. John Gilmore famously said that the 'Net interprets censorship as damage and routes around it.

But surely there were other ideas worth considering that wouldn't have demanded a surrender to censorship. If you had to spend a fraction of a penny each time to post, for instance, even that small commitment would have lent the early 'Net a touch of gravity. A bit of a sense of skin in the game; maturity. Maybe that would have changed the moral climate.

I do genuinely love Barlow, Mitch, and John. We'll find a way through these times, together.

In fact, Barlow and I, and Mitch much more, were involved in a later company, founded by Philip Rosedale, called Second Life, that pointed to a way forward.

For those who haven't seen it, it's an onscreen virtual world with avatars that is optimized for use on a PC or Mac. (It was a big deal before smartphones.) In Second Life, people create, buy, and sell virtual stuff like avatar designs and virtual furniture for virtual houses, so an economy grew within it.

I'm not saying it was perfect, but here was a fairly large example of people buying and selling their bits. Why couldn't that have happened on social media? Would a little gravity, a little skin in the game, have softened the politics of the Internet?

Second Life was yet another example of a phenomenon noted earlier in the discussion of Kinect hacks. A virtual world that was realized in 3-D graphics and avatars motivated an experiment with an economy that valued individual worth. The data, the inside of the computer, felt less abstract, and that made it easier to see through the bits to the people.

Meanwhile, "cyberspace" implementations, like social media, motivate two-tier schemes where ordinary people barter, while the proprietors earn real megamoney from so-called advertisers.* And that pattern has spawned

---

* "Advertising" isn't an accurate term for most of what is called advertising online. You can read why I feel this way in appendix 3.

the largest and fastest fortunes in history, contributing to a crisis in wealth concentration that has destabilized much of the developed world.

Just as I pointed out earlier regarding Kinect data, experience with virtual reality data tends to result in a healthier approach to the Information Age than schemes that are merely informed by VR as a metaphor. Virtual is better than virtual virtual.

Forty-fifth VR Definition: A person-centered, experiential formulation of digital technology that hopefully inspires digital economies in which the real people who are the sources of value aren't ignored.

Silicon Valley continues to believe in bits too much. There's a lot of serious talk about offering ordinary people, consumers, faux-immortality in VR. Ray Kurzweil has promoted the idea. Meanwhile, some of the captains of Silicon Valley invest in often nutty schemes to achieve physical, biological immortality for themselves.

Birth of a Religion

The worst offender when it comes to believing in bits too much is artificial intelligence. AI's talking points were codified during the period chronicled here. My friends and I had marvelous arguments about it.

I used to ridicule AI when people only talked about it *as if* it were a religion, but a threshold has been crossed: AI is now held as a tender belief by those who hope it will grant immortality, reunite the departed, or solve all of mankind's problems automatically; govern over us all with infinite wisdom. It's been an amazing experience to watch the birth of a religion.*

Because I saw it happen, I understand AI as an old, innocent little thought experiment that turned into effective storytelling for fund-raising, and eventually exploded into a screwy belief system that makes its own advances less useful.† But at the same time, since AI beliefs have become

---

* It's made me reimagine how some of the ancient religions might have been born. Is a second generation of believers always more rigid than the first?
† Not unlike the way extremophilic approaches to earlier ideas for organizing human affairs make those ideas less useful. For instance, markets are useful, but extreme libertarian beliefs along the

sincere, I must switch my discussion into the mode of religious freedom and tolerance.

Religious tolerance has to go both ways, however. I don't believe bits mean anything except as interpreted by people, and I want the freedom to at least explore implementations of a future society based on that premise. But some AI believers have become fanatics, unable to even ponder that another point of view could exist; tolerance isn't even on the table.

Back to the Little Hunan, circa 1990.

From a hardware engineer, popping dumplings, "Okay, look, Jaron, I'm with you on this AI thing. AI scares the crap out of me. I have nightmares about the computers suddenly evolving and just eating humanity, just whisking us out of the way. I don't believe any of the arguments about how they'd be fond of us or keep us as pets."*

> *"Oh god, that is NOT agreeing with me about AI. Fearing AI is even worse than loving AI. When you get scared of AI, that's when you really believe in it the most. If you get people scared of the devil, that's not only the most intense promotion of religion, but it's the flavor of promotion that probably does the most to make religion intolerant. When people get scared, they get narrow-minded."*

"Devils aren't real, but computers are real."

> *"What if AI is just a fantasy we see in the bits we set? What if it's a way of avoiding human responsibility?"*

"This has been argued for decades. When people can't tell AI from people, then AI will be real. You know, the Turing Test."

From a bearded hacker who has been too entangled with noodles to speak until now, "He's got an answer for that."

> *"Yeah, I do. You think that people are these fixed quantities waiting for AI to catch up and then surpass us. But what if people are dynamic, maybe even more dynamic than computers? What if you change yourself when*

---

lines that markets should be the *only* organizing principle in human affairs—or that an unregulated market will always tend toward perfection—have the effect of making markets less useful. Similarly, democracy is useful, but a determination that every little decision should be undertaken through a democratic process as frequently as possible makes democracy less useful. As for religion, don't get me started.

* Stephen Hawking and Elon Musk are the primary current public faces of this fear.

you're around computers? What if you make yourself dumb to make a computer look smart?"

"That could never happen."

Present-day Jaron must interject, from this doubly indented location,* to defend his old self with more recent stories. It has happened! Now that computation runs our lives, we make ourselves dumb to make computers look smart all the time.

Consider Netflix.

The company claims that its smart algorithm gets to know you and then recommends movies. The company even offered a million-dollar prize for ideas to make the algorithm smarter.

The thing about Netflix, though, is that it doesn't offer a comprehensive catalog, especially of recent, hot releases. If you think of any particular movie, it might not be available for streaming. The recommendation engine is a magician's misdirection, distracting you from the fact that not everything is available.

So is the algorithm intelligent, or are people making themselves somewhat blind and silly in order to make the algorithm seem intelligent? What Netflix has done is admirable, because the whole point of Netflix is to deliver theatrical illusions to you. Bravo!

(And by the way, after decades of self-certain, snide arguments against copyright and for making art and entertainment "free," into an all-volunteer zone, look what happened when companies like Netflix and HBO were able to get people to pay for subscriptions for good TV. Suddenly we're in a renaissance that has been dubbed Peak TV.)

Your friends, lovers, purchases, and insecure gig economy gigs are brought to you by acts of misdirection that echo Netflix's moot algorithm. A bounty of options seems to be out there on the 'Net somewhere, too many to evaluate on your own. Life is short, so you suspend disbelief and trust in the algorithms. A fool is born.

From a sweet but sad mathematician, nursing a won ton soup: "Jaron, you talk about VR as if it's the opposite of AI. But aren't they going to

---

* Hopefully this doesn't make my ideas marginal.

converge? I mean, if you look at Moore's Law, we should be able to calculate the year when VR sex will get better than real sex. Algorithms will get to know you and automatically design the ultimate partners for you. I did some preliminary calculations, and I think it might happen in 2025."

This idea has come to be known as the "sexual singularity."* You could spend the whole day unpacking it. May I leave that as an exercise for the reader? Here I'll recall only one narrow line of response that was typical from me:

> *"You're thinking about this backward. It isn't about what an algorithm can do for you, but about whether you can expand your mind. At the end of the day, that's all that computers can help us do. Why not think about sex as something you can get better at yourself? Then you'd not only be connecting with another person, but you'd be alive, growing, changing. Not stuck in a loop with an algorithm. If a device calculates a perfect sexual experience for you, then what's really happened is that you've been perfectly trained in a Skinner box. Don't be a lab rat."*
>
> My answer was similar when people told me that AI algorithms would someday compose ideal music, write ideal books, or direct ideal movies. The premise is backward from the start.

"But what if people like it? You're acting all superior, but what if someone out there likes automatically designed virtual sex partners and books and music that have been created by algorithms to be perfect for them? You're judging! We're all entitled to our tastes."

> *"We're committed to being good engineers, right? What I'm saying is that we screw up the feedback loop that lets us be good engineers when we treat machines as people."*

"You're making a simple issue complicated."

> *"No, think about it. When you grant another human the faith of personhood, the belief that the other person is really a person, then deference comes into play. You can't go around redesigning people. That's fascism. You have to let people invent themselves, even if they're annoying. And*

---

* The sexual singularity is the hypothetical future moment after which virtual reality sex would be more appealing than real sex, and, according to the typical framing, women would lose their power over men.

they often are, but that's what we love about humanity, right? The open-ended unpredictability, the diversity. If you decide to treat computers as people, then you offer computers the same deference. That means you lose the grounding that anchors design decisions about computers. You can't make them better anymore."

"Maybe we need to stop being squeamish and accept that we need to redesign people."

"Oh god no."

"I don't see how you can build a noosphere without accepting that people will have to be redesigned. At least a little!"

"I don't feel this imperative to make a noosphere as quickly as possible. What's wrong with the Star Trek *idea of the Prime Directive?* Let civilizations emerge on their own terms without a top-down designer. Then you get depth and diversity. What's the rush?"*

"That makes it sound like we're superior aliens deciding what to do about Earth." Murmurs of approval around the table. Maybe that's what we were.

From the rail-thin fellow picking at the tiny bones of a whole fish, sensing that it was time to redirect the conversation: "You can only say AI isn't real because it isn't real *yet*. When it's working, the evidence will be overwhelming."

From the beard, morphing into noodles: "Watch out, you're about to get whacked by Jaron's 'premature mystery reduction' rant."

"I guess I've tortured you enough with that one. Can we just agree that all of us might be surprised by how things develop and that none of us should be so sure we know what will happen?"

"It's axiomatic. Moore's Law tells us computers will get millions of times more powerful and then do it again, and then again. They'll surpass our brains. They'll deserve rights. They'll demand rights."

"What you're really asking for is that I bludgeon you again with my circle of empathy argument. About how if you make your circle of empathy too

---

* The famous prime directive in *Star Trek's* fictional universe was to not interfere in the affairs of less technologically advanced planets.

wide, you become incompetent and don't help anyone. You make yourself ridiculous when you prop up perfect little beings, you know, the ones I call 'specialness surrogates.' Like when, um, certain people stopped brushing their teeth because they didn't want to kill bacteria."

Uncomfortable rumblings around the table.

"You're not arguing fair, bringing that up as an example. What I think you're really upset about is how you were going to get sued for virtual fetal support."

> *Present-day Jaron interjects: "We'll get to that story in the next chapter."*

1990 Jaron said, "I wish you guys would at least think about how we must sound to nontechnical people out there in the world. Can you imagine hearing wizards talking about how they'll create superior forms of life, and old-fashioned people will either become obsolete or be kept as pets or something? Don't you think that'll make them just mistrust the modern world? Won't they hate us? Won't they become easy marks for charlatans who convince them science isn't a friend to them? Isn't the whole point of engineering that we're supposed to serve nontechnical people?"

> Please keep the following in mind when you read "think pieces" about how robots deserve empathy: Tech writers have a bad habit of articulating "big ideas" that happen to serve the interests of the big tech companies at a given moment. There were a lot of pieces about the evils of copyright when Google was making an unprecedented instant fortune by plowing over copyright. Similarly, a flood of "radical" think pieces praising the end of privacy and the value of collectivity appeared when Facebook was first commoditizing and cornering the market on digital personal identity.*

---

* An extreme example of nutso fealty to software is the way corporations have become people, at least according to the U.S. Supreme Court, at the same time that corporations have become algorithms.

The algorithms running companies like Google and Facebook are among the only bits that have never been hacked, because they are the only definitive assets in the New Economy. All other bits are somebody else's problem, but the algorithms are seriously protected.

After all the talk about open source and sharing, the algorithms are the only successfully kept secrets left on the planet. Every other asset—the content, in other words—is provided by third parties so that these companies can avoid responsibilities and liabilities.

"If, well, *when* giant robots or superintelligent swarms of nanoparticles decide you're not worth keeping around, it won't matter what you think. You'll be snuffed out. Then you won't be able to go on about whether they're real or not."

"Now you're pissing me off. Remember that guinea pig that operated the tank with the flamethrower in a Survival Research performance? It was immaterial whether the animal knew what it was doing. In order to see the show you had to sign a contract promising not to sue the people who put the guinea pig in the driver's seat, not the creature itself.

The *only* difference between perceiving an evil AI machine that destroys humanity and perceiving total incompetence on the part of technologists and the military is that the second interpretation is actionable.

Every time you believe in AI, you are reducing your belief in human agency and value. You are undoing yourself and everyone else.

Forty-sixth VR Definition: VR = −AI
(VR is the inverse of AI).*

I know how this must read. The VR guy says VR is the best approach to digital information. He dismisses AI, social media, even the World Wide Web! Isn't he just telling us that what he works on is the best? Doesn't everyone say that?

I'm only human. Can't claim to be free of bias.

My argument that VR is the clearest-minded approach to digital technology is related to the reason why professional magicians make the best debunkers. From Houdini to Penn and Teller to the

---

So doesn't the combination of Supreme Court rulings together with New Economy conventions mean that America has already backed into declaring algorithms to not only be people, but superhuman? Whether or not we realize we've done it?

* A related formulation is that AI is like VR, but with time and space exchanged. That is, an avatar in VR is a spatial modification of a person that still reacts in real time. For instance, a person might appear to turn into a lobster but is still interacting with other people and everything else about the environment in real time. When it comes to AI, however, data is gathered from people and then mapped and replayed through supposed AI personalities at a *later* time. An AI entity is a *non-real-time avatar,* and that is what allows for the illusion that it isn't an avatar at all. The people from whom the data was taken are no longer in the room when an AI program is run, so it's easy to imagine that the AI program is a free-standing personage instead of a reflection of human data, capital, and agency.

Amazing Randi, magicians are fantastic at seeing through hokum and exposing it. Similarly, the MythBusters (Adam Savage and Jamie Hyneman) were cinematic special effects experts. People who make a living creating illusions know illusions.

VR scientists are the illusionists of science; we're honest when we tell you we're fooling you, and you should take us seriously when we point out that we're not the only ones.

Forty-seventh VR Definition: The science
of comprehensive illusion.

Love the Work, Not the Myth

Don't get me wrong. The lab where I do my research happens to be the premier artificial intelligence lab in the world. I'm superproud of what we accomplish.

And yet I don't "believe" in AI. I wish we used different terms and fantasies to frame the work. The actual stuff we do, the math and algorithms, neuroscience, cloud architecture, sensors, and actuators—all that is amazing, beneficial, even vital for the future of our species. But I believe it would all get better still if we packaged our accomplishments differently.

My conclusion is not shared by many of my colleagues. It is also not shared by most of the tech press, or many of our stakeholders . . . Please realize I am not presenting a consensus opinion here.

Many of my colleagues think of AI as something we build, while I think it is wrapping paper that we put on what we build. The difference can be meaningful or not, depending on circumstances.

If a program is supposed to be like a simulated person you talk to, and that fantasy is the goal, then obviously there's no choice but to conceive of it as AI.

However, if there's a goal *aside* from the fantasy, such as, say, to make analyzing medical records more efficient, then I always argue that we should try to separate out the algorithms relevant to the task—analyzing medical records—and see if we can design a user interface that makes the results as clear as possible, without bringing in imaginary beings. In my experience, when we do that, it's slow, hard work, but results often improve.

VR is amazing at conveying complexity with lucidity. (Recall the memory palace effect, or the way children can learn to manipulate four-dimensional objects.) So VR is a natural strategy to apply to problems like this.

A further step in this argument: If we *don't* build an improved user interface to see the results of our analysis, how can we possibly know if a robot is conveying the most useful results to us? In other words, if we're relying on the robot to understand what's important, and how to best convey it, how can we know if the robot is any good when there's no other channel we can use to check its work? So I argue that we should attack the advanced UI challenge first, maybe with VR, and only consider AI wrapping paper after that. (This priority reflects the phenotropic ideal, explained in appendix 2: that intermediate results should always be in a comprehensible user interface format.)

I know, it sounds boring to say you're creating a tool to find patterns that we missed before in medical records. It sounds exciting to say we're building a robot to do it. But doesn't it sound even more exciting to say we're using VR to do it?

Whether VR would be better or not, AI makes engineering more confusing, even when the underlying technology is beautiful and needed. Which is our priority? Upholding the fantasy of an imaginary being, or addressing a goal like improved medical record analysis?

Alien Virtual Reality

There's one final cultural legacy of the VPL years that I should mention. It is the only belief in bits that might be even more extreme than AI. If AI has become the new religion, then we are about to encounter the new scholasticism. This is going to get nerdy.

A stratum of technical people have become convinced that we already live in VR. That would be a difference that doesn't make a difference, to use Gregory Bateson's formulation, except that it can become a damaging obsession. Being hacked is the greatest fear of the hacker, and if we're already in VR, then maybe we're vulnerable to a metahacker.

Back in venues like the Little Hunan, and in the question sessions of my talks, this idea would often come up. How could we tell if we're already in VR?

I said different things at different times. The nature of the physical world—the way experiments are repeatable and elegant physical laws have never been violated, so far as we can observe—suggests that if we're in somebody else's VR, the operator doesn't micromanage. I argued that believing in a VR operator was similar to believing in a god, but a primitive type of god with superpowers, not a transcendent god or one with moral significance.

This led to long, long conversations.

Here's another argument I used to make: Why would the god running a VR system in which we live not be embedded in yet another VR system operated by a yet higher god? At the top of a chain of primitive gods would be a more profound idea of God, which is just as accessible without thinking about all the intermediate gods. Ultimate reality is always right in front of you, so why even pay the middlemen any mind?

Or: I predicted that the success of quantum cryptography would suggest we're not being observed, that we're not in VR in any consequential way. Quantum cryptography uses the most fundamental properties of nature to absolutely prove that a message hasn't been read before. The act of observation changes a quantum system no matter who observes, man or god.

Quantum cryptography turned out to work! So, if you believe that argument, there's now less reason to worry that we might be in VR.

Another line of argument is usually associated with the more recent work of philosophers like Nick Bostrom, but it was around back at the Little Hunan.* Roughly: If there is a multitude of alien civilizations, a bunch of them would develop high-quality VR, so there would be many VR systems running, but only one real universe. Therefore, when you find yourself in a reality, the chances are that it's a virtual reality.

The way I used to answer this and similar ideas was that there's probably more than one real universe. Recall Lee Smolin's idea of a landscape of universes that evolve to settle on properties that support interesting chemistry. Since Lee put this idea forward, variants have blossomed. String theory now has its own version of the landscape of universes. There can be

---

* Roboticist Hans Moravec might have been the originator.

an unbounded number of universes in some of these theories, so to compare the number of universes with the number of VRs running within them is not a well-defined problem. You end up comparing infinities. Therefore, once again, stop worrying.

None of this would matter, except that I have known a few young technical people, all male, actually, who have done harm to themselves stressing about these questions. It comes about because of a belief that a VR operator would destroy people or whole universes that were contrary to the operator's self-interest or existence.

This need not imply an evil operator, a bad god, but could be a cyber-Darwinian effect. You see, we might be on the way to implementing a super-AI that would be a godlike hacker, and if our reality doesn't lead to that outcome, it will be mooted, like an animal that fails to pass down its genes. The young men who go down this hole believe that their thoughts might doom them, or our whole universe. They tie themselves in mental knots trying to avoid illegal thoughts.* The symptoms are mostly in the form of bad digestion and sleep deprivation, though there are already rumors of suicides.

The antidote for this type of suffering is to stop thinking so much and engage in genuine physical research with VR. Work on sensors, feel the luscious texture of actual, real reality. Work with real people. You can even leave out the VR part. Just work with real people in reality.

Another antidote is to come up with even more exotic, intriguing theories that aren't self-destructive.

Here's a crazy one that I hatched with Stephon Alexander, a theoretical physicist (and jazz musician) now at Brown University. If the universe teems with intelligent life, then surely aliens would want the most powerful possible computers. To run virtual reality, of course.

Aliens would prefer spacetime topological quantum computers, since that would be the most powerful choice. These are hypothetical computers that tie little knots in time and space to store information and compute with it. Our idea was that enough of this type of computing activity would change the curvature of the universe.

Certainly the idea is nuts, but there aren't any saner ideas around to

---

* I don't want to try to explain this problem in full because it makes no sense, but if you're curious, you can look up "Roko's Basilisk."

address a problem, which is that the cosmological constant, which determines how curved the universe is, is a lot smaller than it ought to be. So why not: Alien computers are unbending the universe. The cosmological constant becomes a measure of how full the cosmic hard drive is, since alien information would be stored nonlocally in the whole visible universe. When we look at the night sky, we not only see a universe teeming with life, but with life that hacks. Alien computers would be dramatic, by the way; spaceships in spherical formations pointing lasers inward to manipulate tiny black holes.*

Alien VR could be doing us good already. Nothing to fear.

---

* Microsoft Research is studying Topological Quantum Computers, but they don't use black holes, so we're not at the point of tweaking the shape of the universe just yet.

MicroCosm

1992 was the year when everything changed. Within VPL the year started just as I pleased, with an exotic torrent of activity.

Wonderful projects: We connected people in Germany, California, and Japan in live, shared, virtual worlds and also got them to inhabit telepresence robots across continents. We slaved a robot hand to an avatar hand with enough grace to pick up surgical tools. Surgical simulation was extended to the brain.

VPL had an ambitious secret project under way in the early 1990s called MicroCosm. This was to be the first self-contained VR system. For a base unit, it had tracking sensors and a PC with all manner of special cards embedded within a pretty, curvy plastic sculpture. It would have sold, complete, for around $75,000, which was a spectacular cost reduction at that time.

The MicroCosm EyePhone could be converted into a handheld stereo viewer, a little like opera glasses. Instead of being always mounted on the head by a ring, it could also be held up in front of the eyes with a handle, so one could go in and out of the virtual world instantly. In that configuration, visual access to the virtual world could be casually shared. Hairdos were not challenged. The handle was also a control device with both sensors and active haptic feedback. The user's other hand, typically the dominant one, could wear a glove.

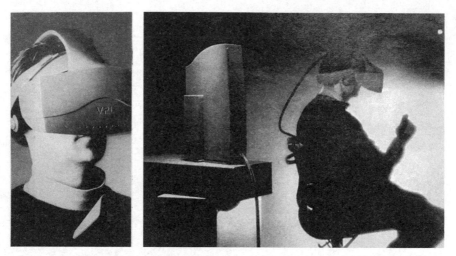

Photos of a prototype of VPL's MicroCosm VR system, which never shipped.
Designed by Ideo for VPL, MicroCosm would have been the first self-contained
VR system. Not shown is the detachable handle with haptic surfaces that
could hold up the headset instead of the head ring, much like opera glasses.
These images are from the October 1992 issue of the design magazine
Metropolis, the only ones ever published. MicroCosm was declared to be the
product that best represented American design, not just of the year, but
ever.

Some of the MicroCosm team testing a prototype. From left to right: Jaron, Ann,
"Comet" Mitch Altman, Dale McGrew, Dave Molici, David Levitt, and Mike Teitel.

Instead of refrigerator-sized computers from Silicon Graphics, Micro-Cosm incorporated some of the first 3-D graphics cards for a PC, designed by our UK partners and distributors, DIVISION. MicroCosm would have shipped in a cute soft case with a handle, and would have been easy to transport and almost instant to set up. Would have.

MicroCosm was wonderful, more usable and comfortable than any VR system I have tried since; also an insanely expensive undertaking for a small company. It never shipped.

VR, Trapped in the Movies

At the start of 1992, we hadn't moved to our fancy new tower with the big octagonal window yet, though we were packing. This was when it started to feel like the real world was getting too weird, that it might break.

There was a drive-in movie theater tenuously plopped on gravel by the bay, a way for the landlords to earn a few bucks until the politics gave way and expensive condos could be built right up to the water. VPL's inland-facing windows framed the faint screens, soaring above couples in cars making out. Usually we saw bouncy car chases and maudlin kisses, but in 1992 we saw ourselves.

The Lawnmower Man was a science fiction film that used real VPL gear as props. It was about a VR company being taken over by a shadowy conspiracy. Pierce Brosnan played, approximately, me.

The movie started out as an adaptation of a Stephen King novel but ended up as a tale of intrigue that was inspired by what might have been the truth about VPL.* I don't really know how true it was, to this day.

According to press reports, our beloved VPL had been targeted by French intelligence services. The French apparently believed we held valuable technological secrets, or at least a French official had hoodwinked his superiors into funding an adventure based on the premise. Supposedly our French investors and board members were connected with a broader covert effort to infiltrate a variety of Silicon Valley companies.

The absurdity was delicious. The French! Infiltrating a virtual reality

---

* According to director Brett Leonard, in conversation.

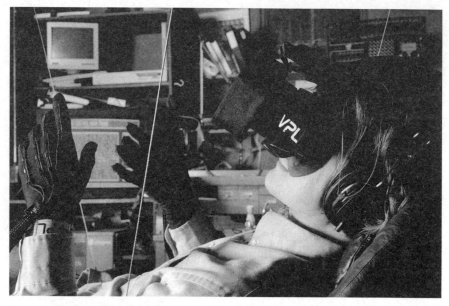

VPL equipment in a still from *The Lawnmower Man*.

company! One newspaper headline read, VIRTUAL REALITY HAS BEEN PENETRATED, and another, VR COMPANY BARELY EXISTS.

The *National Enquirer*, a reliable supermarket tabloid, ran an earnest article revealing that the CIA employed a secret underground bunker's bushel of spies who donned DataGloves. All day, the spies would wear DataGloves and wiggle their fingers, remote controlling severed robotic hands that crawled into enemy territory to snatch papers. This was said to be the prize technology sought by the French.

The French board members were probably not attempting to realize that fantasy, but they were nonetheless unfathomable. (One of them loved using the word "nonetheless" in almost every sentence.)

The board firmly, to the death, refused to allow VPL to go online. Seriously, we weren't able to register vpl.com. It's almost impossible to describe the depth of the absurdity. The stated reason was that it would be a security risk. Someone might be able to hack us and steal files. Also they were worried that everyone would spend their time on the nether reaches of the alt. hierarchy and become grouchy and unproductive.

We gradually became demoralized.

Voluminous Possibilities Lost

VPL felt for a while like one of the majors, but really it still had a mountain to climb and would have had to come up with a means for lurking for decades if it was ever going to be the Apple or the Microsoft of VR. It was way too early.

What we were building was more like the PDP-11 of VR than the Mac of VR. The PDP-11 was a computer every lab needed to have in the 1970s. It was too expensive for regular people, but cheap enough for universities, and it had a great mad scientist vibe, with blinking lights and whirring reels of tape.

VPL was just too small to be a manufacturer and a software company and a cultural force all in one, which does take up time. The core arithmetic didn't work. We were inadequately financed to build expensive VR equipment and then wait forever to get paid. The French board members were obstructionist when it came to helping the company solve that stupid problem.

Other problems were weirder. I started to wonder whether William Gibson was writing our lives.

We got into a dispute with a licensing agency that sat between us and Mattel when the Power Glove came out. There were good people in the firm, and they made real contributions to the product, but the leader was one of those classic New York characters who these days might be called a Trumpish figure.

He was amazing at sales, at working a room, but also crazy emotional. This is the thing about master salesmen; they ultimately fool themselves in order to fool others.

He had a 1980s Hollywood guy power look—longish wavy hair that would have appeared "hippie" if the cut hadn't been manifestly fastidious and expensive—and lots of bling. He'd make every conversation into a rung of the sales process, and every lack of perfect agreement into a hysterical yarn.

"Look at these eyes. You aren't looking. *LOOK*. These are the eyes that didn't look at their own mother for years—sweet mother, you have to understand, a dear lady—these eyes would not even look at her after she stonewalled me. Now you're telling me you don't think I can walk away from this deal?"

It went on like that, and what was amazing is that it often worked for him.

The guy had a self-defeating criminal streak; he'd been convicted of fraud and other crimes in a federal court. You ask: Why did I make a deal with a guy who had a criminal past? The lawyers and the board asked the same question, and in a less polite way than you just did. The answer was obvious. I had no idea I was doing it. Inexperience.

Some years earlier the colorful fellow had decided he wanted to hold back royalties, and we went to court, and then he relented, and the problem evaporated.

In order to settle, he asked only that I agree to attend while he pitched deals to other parties involving VPL technology. I wasn't obliged to accept any of these deals, only to attend. To be a character in a stage show to be raised by master salesmen. Fine.

The time had finally come for me to fulfill this obligation, but the people we were supposed to pitch made no sense. The list included Imelda Marcos, Donald Trump, and Michael Jackson. The guy wanted me to fly around the world full-time for the whole year, basically to be a pawn in surreal shots at implausible deal closings.

It was fun to hang out with Michael Jackson's family in their kitchen. I guess any encounter with Michael was likely to be odd in those years, but the master salesman had hinted that my hair might have cooties, and Michael was obsessive about such things. So when we talked about technology, gathered at a wonderful, huge analog mixing console, I stuck to the submix faders at the extreme right, while he stayed with the input channels at the extreme left, and we politely yelled back and forth in our high voices.

The master salesman never made it through the whole list of targets. I probably cramped his style.

The weirdness of real reality was outrunning the weirdness of VR, and I didn't know how long that could go on. Nothing was making sense.

In Japan, I learned that one of our licensees was mixed up in accusations related to organized crime. While it was a chore to collect all the money we were owed by those folks, they did make the delay interesting by showing me a little of the sybaritic, *allegedly* criminal side of Tokyo. Modern descendants of geisha, personal female attendants, entrancing in their flowing veils of all colors, gleams of gold, in gondolas, floating upon a shimmering pool atop a skyscraper, night lights everywhere, Mount Fuji in the distance at dawn.

Recall that VPL had fundamental patents at the time, for things like avatars and networking people together in a simulation. The patents were pure honey, attracting endless litigation and conflict, but it was so early that everyone was wasting their time. As I stated earlier, the VPL patents have long since expired; only now, decades later, is VR ready to make real money.

The *perception* of value when that value is not actionable is the worst kind of magnet for strife. No one ever comes out satisfied.

In Slickness and in Stealth

The marriage was brief but painful.

She said, "You don't appreciate my facets."

"It's your edges."

"Ha-ha. You're too clever but not clever enough."

The divorce threatened to be an absurdly titanic struggle.

A celebrity Hollywood divorce attorney named Marvin Mitchelson decided to try a new twist on me. The proposition was that I had denied my ex-wife impregnation, and she was old enough that her biological clock was ticking. Therefore the threat was that I would be sued for "virtual child support." What he was really after was a piece of those patents. As it happened, Mitchelson was disbarred and imprisoned for unrelated matters before the suit could progress. (My ex-wife later had a child with a different father, so all ended well.)

But along the way, it felt like I was followed constantly. One morning I set out for a long walk, all over Manhattan, to think. I was still living in California and was only in town for business. When I sat down on a random bench in Central Park in the late afternoon, I was served with a document related to the divorce. The process server must have been told to follow me all day so that I'd know I was always being watched.

This experience clarified the difference between theory and experience. I had long believed that it was both true that human life was mysterious and sacred, AND that one must not commandeer the bodies of others. Thus I supported women's rights to choose abortion.

But I had never experienced the prospect that the law might commandeer my own body in the service of someone else's ideas about reproduction. I was told I'd have to provide a sperm sample so that my virility could be proved in court, for instance. As peculiar as this episode was, I wish other men could experience what it's like. The debate about abortion would be quickly put to rest.

As a side effect, I had the misfortune to know what it's like to be tracked and followed all the time. At that moment, on the park bench, I had a practically heart-stopping revelation: If events unfolded without check, Silicon Valley would soon be following everyone around, just as I had been followed. You can't live that way unless you live in denial, and eventually you have to die a little inside to pull *that* trick off for a lifetime.

Unfortunately, we digital gurus were about to push everyone away from the honest experience of life, for how could anyone ever accommodate living under surveillance unless they also lived in denial?

But at least I was able to enjoy one last, marvelous occasion when I put a MicroCosm prototype to good use.

The Sound of One Hand

In 1992, SIGGRAPH was held in Chicago. The film show is always a highlight of the event, giving the industry its first look at the latest special effects amid whoops and hollers. There's usually a halftime show with a live event onstage, and that year I performed music from inside virtual reality on a MicroCosm. This was the only time a MicroCosm was ever shown in public.

I started on the design of the virtual world, which I called the Sound of One Hand (since I played the music with a single DataGlove), and on learning how to play it, only a month before the performances. I had to become completely immersed in the creation process. What a reckless luxury it was to dive into music. In retrospect I realize I was already experimenting with what it would be like to leave the tech business behind.

In the performance, I used a MicroCosm headset, with the opera handle, to be inside the virtual world, but the audience could see what one of my eyes was seeing on a big projection screen behind me. And of course they could hear the music I was making.

Every note of the Sound of One Hand was generated by my hand movements, as they were transmitted through the DataGlove to the virtual instruments: There were no predetermined sequences. It's not easy to make an audience believe this is true. A performer might just be miming to prerecorded events. To use interactivity onstage, you have to start with a mini-demo phase that convinces the audience that the interactivity is real.

The first virtual instrument I played in the show, in order to demonstrate the interactivity, was called the Rhythm Gimbal. (A gimbal is a common mechanical construction of a hierarchy of rotating joints.)

The Rhythm Gimbal resembled a gyroscope. When it was still, it was completely white and completely silent. When I picked it up and moved it, it began to emit sound. Actually the sound was triggered by the rings rubbing against each other—they also changed color at contact. Once set in motion, the Rhythm Gimbal would slow down, but would take a long time to stop completely. Thus, unless I carefully released it without any spin, it would continue to make sounds when I wasn't looking at it. The "background" sound heard while I played the other instruments came from the Rhythm Gimbal.

A range of harmonic and textural styles could be explored by practicing the spinning of the Gimbal, varying from an open, harmonic, and calm sound to a crazed dissonance. My favorite was the in-between zone, which sounded like a cross between late Scriabin and the Barber Adagio (no kidding).

It was almost appalling how good this simple gizmo was at generating harmony. Is this all a composer's brain does? But the Gimbal can't be described properly as a self-sufficient algorithmic music generator. There is a necessary element of intuitive performance in finding the weird harmonies of this curious instrument.

I couldn't get a specific chord out of the Rhythm Gimbal reliably, but I could get a feel out of a chord progression, because I could influence when chords changed and how radical the changes would be. This did not feel like less control, but a different kind of control. The test of an instrument is not what it can do, but whether you can become infinitely more sensitive to it as you explore and learn. A good instrument has a depth that the body can learn that the verbal/visual mind cannot.

The original plan was that the Sound of One Hand would be an elaborate VR "demo," or explication. But as I worked on the world, a mood, or an essence, started to emerge, and furthermore it was true to my emotional and spiritual experience at the time. This was unexpected and exciting, even if the content was not cheerful. So I went with a darker and more intuitive process instead of falling in line with the more familiar computer culture of clarity and light humor. There have only been rare occasions when I felt I was programming in an intuitive way—it's not easy to bring one's technical and emotional capacities into alignment—but this was one of those times.

There were other instruments,* all floating inside a hollow asteroid, me flying around them, lost, alone, making music for an audience unseen.

Computer music, because of its coded nature, can't help but use instruments that are built out of concepts of what music is. This is a drastic departure from the "dumb" instruments of the past. A piano doesn't know what a note is, it just vibrates when struck. A sensitivity, and a sense of awe, at the mystery that surrounds life is at the heart of both science and art, but instruments with mandatory concepts built in can dull this sensitivity. If you pretend that what you can code reflects a complete understanding of what you can do, then you lose sight of the mystery at the edges of everything.† This can lead to "nerdy" or bland art. In order for computer art, or music, to work, you have to be extra careful to put people and human contact at the center of attention.

I was delighted to discover that the Sound of One Hand created an unusual status relationship between the performer, the audience, and the technology. The usual use of rare and expensive high technology in performance is to create a spectacle that elevates the status of the performer. The performer is made relatively invulnerable, while the audience is supposed to be awestruck.

The Sound of One Hand set up quite a different situation. The audience watched me contort myself in all manners as I navigated the space

---

* The Cybersax was the most ergonomically complex instrument. When the sax was grabbed, it turned to gradually becoming held by my hand, and it tried to avoid passing through fingers on the way. Once I was holding it, the positions of my virtual fingers continued to respond to my physical ones but were adjusted to be properly placed on the sax keys. This is an example of the filtering of control that is critical in the design of virtual hand tools, especially when force feedback is not available.

† This use of the word "everything" is the same as when Leonard Cohen's "Anthem" refers to the "crack in everything; that's how the light gets in."

and handled the virtual instruments, but I was wearing EyePhones. Five thousand people watched me, as I took on awkward poses, but I couldn't see them, or know what I looked like to them. I was vulnerable, and very human, despite the technology. This created a more authentic setting for music. If you have played music, especially improvised music, in front of an audience, you know the vulnerability I am talking about, that precedes an authentic performance.*

The Sound of One Hand was a bigger leap into the unknown than all of the weird "experimental" performances I had been involved in New York in the late seventies. I had absolutely no idea if the piece would take on a mood or a meaning or if the audience would find the experience comprehensible. The performance turned out to be a cheerful, therapeutic event for me. It was a technological blues, a bleak work that I could play happily. It was a chance to work on a purely creative project with the VPL family, a chance to treat all of VPL's stuff as a given set of (reliable!) raw materials instead of as work to do, a chance to practice what I preach about virtual tool design, a chance to use VR just for beauty, and a chance to be musical in front of my ridiculously ambitious professional peer community.

The audiences were incredibly responsive, and I didn't hear anyone describe the piece as a demo. It was music.†

The End of One Finite Game

In retrospect, the Sound of One Hand was my *Fitzcarraldo* moment. After mountainous work, a performance, and then off into the wilderness.

SIGGRAPH '92 was a high, but afterward I had to face reality. I was not seeing eye to eye with the rest of VPL's board. We were stretched thin, it was true. I wanted us to go all out for MicroCosm, even if it meant risking the whole company. And I wanted VPL to pounce on the growth of networking. We had software that was already primed to become one of the first networked apps. The other members of the board wanted VPL to change course to become a less risky, more conventional company that had military contracts, sold a low volume of high-cost, high-margin items, and waited until

---

* Writing a book sets up a similar power dynamic, I have found.

† www.jaronlanier.com/dawn

the intellectual property was worth enough to sell at a high price. The board's vision came with an extraordinary price: strategic bankruptcy.

According to this plan, VPL would be taken through a bankruptcy, after which a new version would emerge, free of creditors—who I must point out were also board members, but never mind—and then would be thoroughly controlled by the French investors. This is when I could have fought and perhaps won. I saw an opportunity to make a high-stakes bet, but no need for a bankruptcy.

In the back of my mind I was starting to wonder if I had started to chase the wrong dream. If I had wanted to be one of the Silicon Valley moguls, I still had my chance. There was a path to making VPL into one of the big companies, and it had to do with the rise of networking. I could possibly fight the various ridiculous battles with an ineffectual board, the alleged Japanese mob, alleged French spies, and confirmed Hollywood divorce attorneys and come out on the other side leading a big tech company. But I had fresh doubts about whether it was what I wanted.

If I was to make an uncharitable assessment of myself at the time, I would say that my problem was that I wanted to be loved. I was still that little boy who had lost his mother, and I couldn't bear the jealousy and pestering that came with success in Silicon Valley.

But that wasn't the whole story. I also doubted the mythical dimension of masculine success in the business world.

There's a vibe in the business world that is a faint echo of military culture. A leader is someone who can magically will events into being. Steve Jobs talked about "denting the universe." Just as New Age superstition held that one's thoughts created reality, so it was in the masculine mythology of business. Magical thinking, over and over.

And yet I had known a number of businesses intimately by this time. Tech companies, giant toy companies, military contractors. So far as I could tell, what actually happened didn't correspond to the mythology of the business superman. Leaders struggled with one another over power and reputation, but whenever anything actually got done in a useful way, it was thanks to an uncelebrated person, an invisible angel. While my myth-making was more visible, people like Chuck and Ann did as much or more to get VPL to accomplish anything substantial.

If you cease to believe in the myth of the great man, it's hard to aspire to be one.

Thus I came to an incomprehensible conclusion. It was time for me to leave VPL.

It was like renouncing your country or your religion. I was disoriented and unsure.

VPL went on without me, selling the same wares, not innovating, so far as I knew. I didn't follow it from afar. It was acquired in 1999 by Sun Microsystems, which eventually became part of Oracle.

Again, a part of me died, and it was time to start anew, forgetting as much as I could bear.

21. Coda: Reality's Foil

My years since 1992 have been filled with every kind of wonder. I've changed; everything's changed.

I'm surrounded by the sweetest possible family. I'm happy.

As for the far world out there, the story is mixed.

El Paso had been a terrifying place to me when I was a child, but today it's one of the safest cities in America. It feels less ethnically divided and more relaxed; the spicy bubbling of its cultures is pleasant and closer to the surface.

Meanwhile my sweet Ciudad Juárez became known for a while as the murder capital of the world. Young women were gruesomely disappeared en masse. Walking over the bridge from El Paso during the darkest years, from about 2008 to 2011, felt like descending into a medieval hell. Now the city seems to be crawling back out of the darkness.

In other news, patrons no longer smoke in American restaurants. A young version of me can play in a restaurant band.

New York City and Los Angeles achieved reasonably breathable air. Visiting big American cities no longer feels like landing in the atmospheres of alien planets.

However, Manhattan is overrun by the same chain stores you find anywhere. It feels less exceptional than it used to.

At the same time, cultural depth took root in L.A.; the vague city of

A 2015 gathering, at Microsoft Research's Faculty Summit, that included many of the VR researchers mentioned in this book. This is not a photograph, but a 2-D image of a volumetric capture. From left to right: research intern Victor Mateevitsi; Ken Perlin, NYU; Mark Bolas (in a HoloLens), then at USC, now at Microsoft; research intern Andrea Won; Christoph Rhemann, Microsoft; Andy van Dam, Brown University; me, Jaron; David Kim; Henry Fuchs, UNC; research intern Joseph Menke; Steve Feiner, Columbia; Shahram Izadi, Microsoft; Blair MacIntyre, Georgia Tech; Carolina Cruz-Neira, University of Arkansas, in a HoloLens; research intern Kishore Rathinavel; Tom Furness, University of Washington; research Intern Gheric Speiginer (practically swallowed by a Reality Masher experimental headset); Ken Salisbury, Stanford; Wayne Chang, Microsoft; Jianxiong Xiao, Princeton; Ran Gal, Microsoft; two unidentified visitors; Javier Porras Luraschi, Microsoft; Zhengyou Zhang, Microsoft.

indecipherable houses no longer feels like a dead end. So far as I can tell, it brims with people living substantial, fulfilling lives. What changed more, L.A. or my ability to perceive L.A.?*

---

* It feels like breaking a mutually flattering vow of silence—we're all supposed to know that each other knows—but I must mention one of the most basic qualities of this book, and of storytelling in general. One of the reasons I've taken the time to describe how various social scenes and physical environments seemed completely different to me at different times is that I want to emphasize the reality of inner life. A feeling is not a fact. You might have strong reactions to something you perceive, but once in a while those reactions might have more to do with your internal mental and emotional processes than with the nature of the thing out there that you are perceiving. Digital culture has become so centered on behavior and measurement that we can easily forget that the way the world feels to us might not just be about the world, but about how inner experience colors perception. Maybe if you get upset about vaccines, gluten, the way women sound when they talk, immigrants, political correctness, or whatever it is that makes you boil . . . maybe it's not entirely about that stuff. Maybe it's worth noticing what's going on inside. I have found that increased

Silicon Valley has changed most of all. We won! We control rent-a-mom. Boards don't meddle. Hackers own companies outright.

We told the world to change for our pleasure and it did! Kids all over the globe gave their privacy over to us, and our algorithms now twitch their marionette strings. They peck against the buttons of our Skinner boxes.

Hackers own the wealthiest companies the world has ever seen, but have relatively few employees to worry about. Individual young hackers routinely and instantly become richer than significant fractions of the rest of the world's population.

Kingdoms have cycled. Sun's old campus is now Facebook headquarters. Silicon Graphics' old headquarters is now the Googleplex. (I used to do VR experiments in what is now the Googleplex lunchroom.)

What about the *feel* of being in the Valley? One big change is ethnic diversity. A typical meeting these days includes lots of engineers from India and China and everywhere else.

My sense is that there's been a *slight* lessening of cognitive diversity, however. Everyone seems to be reflecting at least a little more energy on the autism spectrum than in the old days.

Another change is political. The Valley is still fairly lefty-progressive, but a libertarian strain has become quite intense.*

At the start of this book, my younger self thought the future sounded like both hell and heaven. There's certainly plenty of hell to go around lately.

---

awareness of inner life, of experience, makes both me and other people into more effective scientists and engineers, and also sweeter.

* I can understand why many younger techies have swung libertarian. The aspect of Silicon Valley run by government can seem as if it was designed to frustrate techies. It's a frequent complaint.

For instance, our Bay Bridge was originally built between 1933 and 1936, and was the longest in the world.

The bridge is actually two bridges that connect through a tunnel in an island in the middle of the bay. One portion was damaged in a 1989 earthquake. The damage wasn't in the glorious suspension bridge on the San Francisco side; that ancient engineering held up. The problem was in the low-rent span leading to Oakland.

There were so many meetings that repairs didn't start until 2002; the replacement span wasn't opened until 2013. The demolition of the damaged section won't be complete until after this book comes out.

If you're a millennial in Silicon Valley, the government has taken longer than your whole life to fix a bridge that took only three years to build in the primitive conditions that predated even the most experimental early computers. Meanwhile, Silicon Valley products like the iPhone and Facebook can change the world in a few months.

I believe democracy is worth it because I've seen enough of the alternative, but if all you've ever seen is tech companies and the Bay Bridge, you might feel differently.

Digital hothouses of irritable politics migrated from Usenet's alt. hierarchy to Reddit, 4chan, and other hubs, and nurtured outbreaks of ill will like Gamergate, and most recently the alt-right. Unfortunately, the story of VR is intertwined with that migration. Read the sorry tale in appendix 3.

Despite the long history of inadequacy in cautionary tales about computation—people always seem to *want* dystopian tech because it looks so cool*—I tried writing in the genre myself, starting right after I left VPL.

I wrote essays about how society might someday be made absurd by abstract wars between algorithms, and how "viral" online dynamics could enable sudden social and political catastrophes. My cautionary tales were nicely appreciated in some circles, but obviously didn't forestall the events I warned against.

Here I am, trying again, but with a difference. Enough has happened that the tales are no longer cautionary. I'm laying down breadcrumbs for you to follow, to get a sense of how we got to where we are. Will it be helpful? I hope so.

But let's turn to more pleasant topics.

I still love VR. I just do. It's such a treat to try out virtual worlds coming from young designers. It's delightful to watch people have giddy VR experiences.

VR still teaches me. I love noticing my own nervous system in operation, and that's more possible in VR than in any other circumstance. I love seeing nuances I used to be blind to in the light and motion of the natural world, in forest leaves† and in children's skin. This happens most intensely when you compare reality and VR.

The science of making VR gear better is still fresh and new. It's still a thrill to find a better way to pick up a virtual object.

---

* The hit of the 2017 Consumer Electronics Show was a stationary round device that listens, talks to you, and optimizes your life. People want to buy HAL!

† While I was writing this book, a student intern of mine, Judith Amores, produced a HoloLens app for adding art and sculpture to the real world. You can sculpt rainbow structures, spray graffiti on walls, or even people, and it sticks, or you can lob gobs of goop, which splat on the world, then sag and droop.

I took it to the forest. You can easily just walk by a tree and take it for granted. You can deface or destroy a tree, seeing only yourself. Or, you can ornament a tree virtually and then remove what you did. Now the reality of the tree pops out.

Aside from all that, the obvious core joy is that VR can be manifestly beautiful.

What I love most is watching other people love VR. Virtual reality had another revival in the mid-twenty-teens. A new generation not only discovered the joys of VR, but became fanatical.

Sometimes I'm asked if I'm annoyed when a twentysomething seems to think VR was invented only a few years ago, or only became good enough to be worth a mention when the latest company was funded. I'm not annoyed at all. I'm thrilled. They care enough to want to own it.

Young people *should* own VR. They *do* own VR. Whatever I say is not going to be as important as what the generations to come make of it.

This book has been mostly about "classical" VR, but mixed reality has soared recently, largely because of the development of HoloLens. I'm *so* looking forward to trying virtual stuff that young designers come up with in mixed reality.

How are VR and MR related? They overlap. Future devices might be able to function in either mode. Even if that happens, I suspect VR and MR will remain culturally distinct, in the same way that movies and TV have remained distinct even though they are now delivered through the same channels to the same screens.

By the time this book is released, it's possible the latest wave of classical VR enthusiasm will have crested. But if that happens and you're a young person who has become entranced by VR in the last few years, please realize that there will be more VR waves, and before too long. VR is hard to do well even in a lab, and there's still a lot to learn about how to make great VR products. Be patient.*

I realize I might seem schizophrenic to some readers. If you're a techie, you might wonder how I can spend so much time on downer warnings

---

* We were all spoiled by the way the iPhone took off. It was *huge* right away, and the smartphone genre it launched just got huger over time. But that's unusual.

   It almost never happens. Personal computers didn't blossom so smoothly and instantly, and neither did social media, and yet these designs also became huge. It just took a little longer. Just because it takes a while to figure a technology out, that doesn't mean the world has rejected it.

   But I'm only guessing that comforting thought will be the right thing to say by the time this book comes out. Maybe VR will be huge, huge, huge, and the most important thing will be to tamp down unfortunate excesses that spring up.

about how we're turning ourselves into zombies. If you're a humanist book lover, you might wonder how I can be a cheerleader for tech at the same time. It isn't easy to walk this particular tightrope, but we all have to learn to walk it if we're going to survive.*

As I'm writing, the world feels more and more like a dystopian mid-twentieth-century science fiction vision come to life. That genre invariably depicted the technology of the future as irresistibly cool even as it warned us of the perils.

The other day my family visited another family, and while our children frolicked with HoloLenses, looking as delighted as children can possibly look, the parents fretted about the authoritarian turn of the United States. Didn't that originally occur in a Philip K. Dick novel, or was it a deleted scene from *A Clockwork Orange*?

I find hope in watching younger people use technology. The impressions I'll share haven't been verified scientifically, so far as I know, but please give me a bit of slack.

Young people don't seem to be as easily fooled by online fooliganism. They grew up with stupid social media excesses, so they can size them up. Elders were hit hard by the new technology of lying on social media, and often seem to have been transported into a world that is more artificial than anything I've ever dreamed up in VR.

The younger a person is, the more they seem to be finding a course to wise moderation in their use of technology. Generation Xers seem a little more addicted to their social media feeds than millennials,† and kids seem to get bored faster by the endless foliation of self-similar vanities.

Minecraft would please my younger self especially. If you haven't seen it, it's a whimsical, blocky-looking virtual world, originally seen on PC screens, which is constantly redesigned and reprogrammed by the people who use it. One of the most popular digital designs *ever* for children.

Microsoft bought the company that makes Minecraft, and I got to work with the Minecraft crew on tweaking it for VR release. I tested

---

* Walk a tightrope—"on line"—get it?

† There's actually a little preliminary research to back this observation up, though it's too early to call it a trend. Call it a hope. https://www.nytimes.com/2017/01/27/technology/millennial-social media usage.html

designs with my nine-year-old daughter and her friends, and the word has to be "ecstatic." They don't just learn technical skills, they create beauty. It's better than I had dared hope when I was a teenager trying to find the words to convey dreams about a future like this.

Enjoying technology deeply and fully is the best way to own it; to not be owned by it. Dive in.

Afterword

In 2014, I won an award called the Peace Prize of the German Book Trade. I don't think I can do any better at conveying what happened right at that moment than I did in my remarks at the award ceremony, so I'll end this book by quoting from the conclusion of those remarks:

> It is okay to believe that people are special, in the sense that people are something more than machines or algorithms. This proposition can lead to crude mocking arguments in tech circles, and really there's no absolute way to prove it's correct.
>
> We believe in ourselves and each other only on faith. It is a more pragmatic faith than the traditional belief in God. It leads to a fairer and more sustainable economy, and better, more accountable technology designs, for instance. (Believing in people is compatible with any belief or lack of belief in God.)
>
> To some techies, a belief in the specialness of people can sound sentimental or religious, and they hate that. But without believing in human specialness, how can a compassionate society be sought?
>
> May I suggest that technologists at least try to pretend to believe in human specialness to see how it feels?
>
> To conclude, I must dedicate this talk to my father, who passed away as I was writing it.

I was overcome with grief. I am an only child, and now no parent is left. All the suffering my parents endured. My father's family suffered so many deaths in pogroms. One of his aunts was mute her whole life, having survived as a girl by staying absolutely silent, hiding under a bed behind her older sister, who was killed by sword. My mother's family, from Vienna, so many lost to the concentration camps. After all that, just little me.

And yet I was soon overcome with an even stronger feeling of gratitude. My father lived into his nineties, and got to know my daughter. They knew and loved each other. They made each other happy.

Death and loss are inevitable, whatever my digital supremacist friends with their immortality laboratories think, even as they proclaim their love for creative destruction. However much we are pierced with suffering over it, in the end death and loss are boring because they are inevitable.

It is the miracles we build, the friendships, the families, the meaning, that are astonishing, interesting, blazingly amazing.

Love creation.

Appendix One:
Postsymbolic Communication
(About the Reveries of One of My Classic VR Talks)

More Transcript

The section titled Transcript documents one of my circa 1980/81 talks that began with a discussion of mountain-sized cephalopods and the experience of early childhood. Naturally, this was the only possible way to introduce the main topic, which was, "How can technology become fascinating enough to become more about meaning than about power, forever?"

The transcript continues:

Suppose it is decades into the twenty-first century and robotics has gotten much more advanced.*

Maybe you'll be able to make a Brobdingnagian, animatronic, aquatic, bejeweled octopus-shaped lodge. Or perhaps someday in the future biological engineering will become capable of yielding a city-sized custom octopus with a sleep chamber for humans.

What we've learned about technology is that certain things might get faster and easier as tech advances, but other things come along that still take as much work as ever. As chips get faster, the effort to make the factories for them gets harder and harder.

So it's reasonable to guess that making mega-octopuses for real is

---

* Ha, as I edit this old transcript, it IS finally that far into the future; still no giant artificial creatures lurking in the bay.

going to be a hassle indefinitely into the future, even if we're not sure what kind of hassle it will be.

And don't forget, you'll probably spend more time on politics than actually working on the project. Even if biological engineering is unregulated in the future, there will still probably be rights to negotiate, or regulations about the uses of land and water for something so big.

But there **are** just a few ways that even an adult can spontaneously realize new things in the world.

If you consider the human body, you'll notice that there are a few special parts of it that move as quickly as thought, and with enough variation to reflect a large variety of thoughts.

Do you know? C'mon . . . The tongue and the fingers!

Fingers can play notes on a piano about as quickly as a pianist can think of them. The fastest piano players are spinning inventive improvisations as fast as people can hear. If you don't believe me, listen to an Art Tatum solo; it becomes overwhelming if you really pay attention.

We have used hands to create every artificial thing, albeit through increasingly long, slow trains of indirection. Hands built fire, which melted iron to make blades, and so on.

At the start of the long tech train are always hands, **but** . . . coordinated by tongues. We speak to plan what to do with our hands.

The earliest memories we retain usually coincide with our earliest experiences of language. In order to appreciate language, you have to start by thinking about haste.

Language is the thing people can do to use a tiny part of physical reality that we have the power to manipulate at the speed of thought—like the tongue—in order to invoke illusions of all the other manipulations of reality that we can only achieve very slowly and with a lot of work.

Language is what we call a "hack" in Silicon Valley culture.

A few flicks of the tongue and excitations of the vocal cords, and you can blurt out "Giant amethyst octopus." Compare that with the decades of work it would probably take to realize the creature in reality, even far in the future.

A symbol is a trick for the sake of efficiency. It lets the brain express thoughts to others about as fast as they are experienced, without all the work of realizing changes to physical reality. Symbolism turns the part of the universe we can control, like the tongue, into an invoker of the rest of the universe, and all possible universes, that we cannot control in haste.

Now consider virtual reality as it might exist in the future.

Imagine that someday there will be user interfaces for creating fresh stuff in VR that works as well, and as quickly, as musical instruments do today. Maybe they will even feel like musical instruments.

Maybe there will be a virtual saxophone-like thing you can pick up in an immersive virtual world. Maybe you'll have to wear special glasses and gloves to see and feel it, or maybe there will be other gadgets that do the trick. Pick it up, learn to finger it and blow, and it will spin out virtual octopus houses and worlds of other fantastic things with the ease and speed that a saxophone can spin out musical notes today.

This will be a new trick in the repertory of the species, a new twist in the human story. The same parts of your body that were used to make language possible will be leveraged to make the stuff of experience, not symbolic references to hypothetical experiences.

True, it will take years to learn how to play things into existence, just as it takes years to learn to speak a language or play the piano. But the payoff will be tangible. Other people will experience what you breathed into being. Your spontaneous inventions will be objectively there, shared to the same degree that perception of a physical object is shared.

In order to approach this ideal destiny, VR would have to include that expressive reality-emitting saxophone or other protean tools, and it is an unknown whether these tools can be created or not. But suppose it can be done.*

Then virtual reality would combine qualities of physical reality, of language, and of innocent imagination, but in a completely new way.

This destiny for virtual reality is what I call postsymbolic communication. Instead of telling a ghost story, you'll make a haunted house.

Virtual reality will be like imagination in that it will engender unbounded variety. It will be like physical reality in that it will be objective and shared. And it will be like language because adults will be able to be expressive with it at a speed that is at least comparable to the speed of thought.

* The second appendix, on phenotropics, explains a little about how I think it might be done.

Kind of Blue

This is where the transcript ends, but I remember what would have come next. Lots of questions. I'll answer just one of them for now.

> *"Wait a second, wouldn't virtual things just be new kinds of symbols? Wouldn't they be abstract or Platonic references to things that might come to be? Are they really that different from words?"*

Glad you asked. The first thing to understand is that we don't yet have scientific descriptions of meaning, symbols, or abstractions. We can't describe these things as phenomena in the brain. We've pondered the meanings of these words for thousands of years, but we still can't build a detector for whatever it is they represent. We pretend we implement them in computer programs, but that's just a stance for marketing and funding.

Even so, I can offer arguments that postsymbolic communication will be distinct from whatever it is that came before.

Consider "blue." A scientist might describe it as a frequency of light that best matches a class of sensors in the retina. But that's not the whole story. We perceive blue in things that aren't blue in that sense, like seas, grass, and music. So what is "blue?"

Imagine a bucket in VR that has every blue thing in it. Put your head in and, like the TARDIS, it is huge inside, every blue thing floating out to the distance. You'd sense the commonality without needing a word for it.

That would be a new kind of blue. A sufficiently fluid concreteness ought to be able to take over at least some of the duties of abstraction.

Another way to put it is that if the whole universe is your body, then talking would be beside the point. You'd just realize what you would otherwise have to describe. (The two previous sentences might be hard for people who haven't tried VR to understand.)

There are scholars who argue that people in ancient times didn't even notice the existence of the color blue until there was a word for it.* It is absent from much of ancient literature. How can we not wonder what we might be missing today? Maybe postsymbolic communication will open our perception wider than words.†

---

* See *Through the Language Glass* by Guy Deutscher.

† I speculated about buckets of blue for years, but today I can cite concrete experiments. Maybe it's time for my old MacGuffin to kick the bucket.

Speak, Tentacles!

Anyone who came to my talks got an earful about cephalopods. I was obsessed with them.

Fancier cephalopods, like the mimic octopus whom we met earlier, can project images onto their skin. They can also change shape to an amazing degree by manipulating the tentacles and raising welts. An octopus might suddenly turn into a fish, which is great camouflage so long as a predator doesn't fancy fish. Cephalopods evolved to be intelligent. They can not only morph, but morph wisely.

There's a famous video, shot by Roger Hanlon of Woods Hole, of a Caribbean octopus turning into coral so effectively that humans, at least, can't see the difference. Another high-performing cephalopod, a variety of cuttlefish, is seen to transform from one sex to another, perhaps to confuse rivals during mating season. Cephalopods can learn to transform into previously unseen things, like chessboards. (Yes, really!)

In machine learning algorithms, like the ones that can tell cats from dogs, we ask large numbers of people to identify what belongs in a category; cats, dogs, or blue things. They usually do this work for us for free as part of an online game or novelty.

Then we use feedback networks of statistical correlations, called machine learning algorithms, to capture what all those people told us. The resulting software can classify dogs, cats, and blue things thereafter, often as well as or better than an average person.

So the bucket of my old thought experiment has been implemented. (I even got to work on early examples of the algorithms.)

In the 1980s, when I was giving VR talks, I had to challenge the primacy of abstractions and symbols, because they were the darlings of academics. But now that machine learning algorithms are not only working reasonably well, but earning the greatest fortunes in history, I have the opposite problem.

Lately, everyone needs to be reminded that just because you can tell cats from dogs, that doesn't mean that all of cognition has been understood.

There is something beyond correlation going on in the human brain. We don't just correlate random new mathematical expressions with old correct ones to detect new correct mathematical expressions, for instance. We *understand* math, but we don't understand what understanding is. There's no scientific description of an idea in a brain at this time. Maybe someday, but not yet. We're capable of forgetting that we don't understand. We confuse ourselves easily.

My friend Blaise Agüera y Arcas (formerly at Microsoft Labs, now at Google) and his colleagues have attempted to run machine learning algorithms in reverse to see if platonic images of dogs or cats might come out. What emerges has to be guided with an artist's touch to make any sense, but it can come out fun and surreal.

We don't know if there are platonic dogs or cats in the human brain. All we know is that different neurons fire when a dog or a cat is seen, but we don't know how or why.

Since I don't know what symbolic communication is, or even if it will still be considered a respectable concept in fifty years, I can't really know what postsymbolic communication might be. All these decades later, I still like the notion of postsymbolic communication because it stresses that we should reach as far as we can, to find what can be new in virtual reality.

If only we could transform like that; we'd be natural avatars. We could morph into anything we thought of. Cephalopod life is far from perfect. To their detriment, cephalopods are born from eggs without a parental bond. They're remarkably smart, but they can't grow a culture across generations. Given a choice, I'd still choose to be human.

But with VR, people might approximately become cephalopods with childhoods—provided we figure out how to make great VR design software.*

Poïesis

Some of the ideas from my old talk can be added to our stable of VR definitions. VR might become, take a deep breath:

Forty-eighth VR Definition: A shared, waking state, intentional, communicative, collaborative dream.

Put another way,

Forty-ninth VR Definition: The technology that extends the intimate magic of earliest childhood into adulthood.

The illusions of childhood would be realized. Practical activity would become numinous. I wondered if our infantile natures might turn out to not be so bad if we were honest about them and created sustainable technologies around them.

The emotions of VR cut deeper than a thirst for novelty. People wrestle within the walls that surround our short lives, so able to imagine, but so limited in our abilities to act. Technology is the pounding of our heads against those walls, and we do at least make dents.

So, another definition:

---

* I described this line of thinking at length at the end of *You Are Not a Gadget*, so I won't reproduce it in full here.

Fiftieth VR Definition: A hint of the experience of life without all the limitations that have always defined personhood.

Fascination Versus Suicide

Here's another question that would usually come up, though if no one else asked it, I would ask it of myself: "What's so important about contemplating the nature of childhood and connecting with people in expanded ways? Isn't this a rather obscure obsession you have?"

My answer was that it's about the survival of our species.

If we keep on the path we're on, we'll eventually destroy ourselves. The more technologically capable we become in the future, the more ways we'll have available to put an end to the human story. The numbers game is against us.*

I often found myself sandwiched in-between technology skeptics and technological utopians. I had to frequently restate that I was unambiguously pro–technological progress. The further back you go in human history, the worse things were. Until quite recently, people had as many children as they could because it was expected that some would not live to see adulthood. Disgusting disease was everywhere, as was hunger, and most people were illiterate and ignorant.

Despite this history, I never argued that science or technology *automatically* make life better; they only create options, wiggle room, that people can use to become more ethical, moral, sensible, and happy. Science

---

* Back in hunter-gatherer days, small gangs or tribes kept each other in check; small numbers of people could harm small numbers of people. Then, with the coming of agriculture, scale was rewarded. Walls went up around cities, and violence was regularized into phalanxes. Large numbers of people could harm large numbers of people. Then martial strategy and innovation ramped up. Moderate numbers of people could harm large numbers of people: the Mongol invasions, the British navy.

But now we face Moore's Law–like effects. Small numbers of people will have ever more numbers of ways of killing large numbers of people. The instrumentation of massive-scale violence is getting cheap; eventually it will be virtually free.

Similarly, it used to take a lot of people to spy on a lot of people, as was the case with the monumental Stasi spying and manipulation empire in the East German state. But it has become possible for small numbers of people to spy on everyone else, and also to block most others from doing the same, since digital networks aren't as equitable as advertised.

and tech have never been sufficient, but only necessary, to any hope of moral or ethical improvement.

One of the perennial tropes of the Silicon Valley utopian conference circuit is "abundance." What that word means in our context is that humanity will soon get so good at technology that every human being will be able to live well, maybe even live forever, virtually for free. The idea is sometimes brought up as a rebuke to concerns about our extreme wealth concentration. "Soon, everything you want will be virtually free, so it won't matter who has money."

But humanity has *already* achieved that potential. We got there some-time in the twentieth century. We've become used to the fact that we already have the means to feed, house, and educate everyone. Everyone! And yet we haven't done so. It has long been the dark shame at the core of the whole enterprise of technology.

I argued that technological improvement as a dominant guiding principle would elevate us until we'd inevitably reach a precipice and fall off into a pit of self-destruction. And yet we can't turn back from techno-logical progress because it would be too cruel. But maybe we can recon-ceive and improve the ramp of progress we are ascending as our technologies improve. Maybe there's more than one kind of ramp.

The reason I've been telling you about toddlers, cephalopods, and fan-tastical experiences is that they point the way to a better ramp of progress. A survivable ramp. I called this better ramp the McLuhan* ramp.

Consider that people have been innovating ways of connecting with each other since the dawn of the species. From spoken language tens of thousands of years ago, to written language thousands of years ago, to printed language hundreds of years ago, to photography, recording, cinema, computing, networking; then to virtual reality, and eventually to what I hoped my talk might provide a glimpse of: postsymbolic communication—and then on to what I could not imagine.[†]

A McLuhan ramp is made of inventions, but the inventions don't just achieve practical tasks; they foster new dimensions of personhood—

---

* In honor of Marshall McLuhan, a celebrity intellectual who thrived in the 1960s. He pioneered the study of media.

† William Bricken, onetime principal scientist at the University of Washington's Human Interface Technology Lab, the VR lab started by Tom Furness and birthplace of the first virtual Seattle, has undertaken an exploration of postsymbolic approaches to mathematics. Read about it in his forth-coming book *Iconic Mathematics*.

potentially even empathy. I used to talk about how "the frontier between us" was by definition endless, unlike other frontiers, because we'd become more elaborate through its exploration.

"These dreamy pursuits," I exclaimed, "these mad projects I have devoted such energy to—the philosophical notion of postsymbolic communication; the engineering project of phenotropic architectures*—these are attempts to take tiny steps up on the McLuhan ramp."

In addition to exploring distant star systems, we might also imagine that in the future we'll find ways to know each other better. Since we're fundamentally creative, that process would never end. We'd become more interesting as we become more known.

I would sometimes call it the empathy ramp. As we ascend, empathy would have more and more of a chance.

Fifty-first VR Definition: The medium that can put you in someone else's shoes; hopefully a path to increased empathy.

A McLuhan ramp is different from an achievement ramp—it might not lead to a precipice. It might just keep on ascending. At a certain point, weapons can't get any more sophisticated. Once each person can kill everyone else on the planet on a whim, then the ramp of martial improvement will be complete. The precipice will have been reached.

At this point in my talk I'd usually mention a book called *Finite and Infinite Games* by James P. Carse. It proposes that certain games come to an end and others open up into an endless adventure. A single basketball game ends, but the overall world and culture of basketball need not. Which type of game is technology?

My lectures often concluded with an admonition that "Technologists have a responsibility to come up with media technologies of such beauty, fascination, and depth that mankind will be seduced away from mass suicide."

I delivered that line with force—audiences gasped—and yet it was the flowery hippie mysticism that people remembered. I was going to end

---

* Read about 'em in appendix 2!

my first book, *You Are Not a Gadget,* with that line about seducing mankind out of mass suicide, but my agent at the time insisted it was such a bummer that it would destroy the book's reception.

But truly, dark realism is the only decent foundation for flowery optimism. Be tactically pessimistic and strategically optimistic.

Rue Goo

I mostly stopped giving "guru talks" after 1992, not because of a fading dream, but because people responded too well. It was getting uncomfortable. What finally turned me off was a particularly saccharine event at the Learning Annex that drew an absurdly sycophantic crowd. I didn't want to become an actual guru, even though so many people are looking for one.

Appendix Two:
Phenotropic Fevers (About VR Software)

Mandatory Metamorphosis

This is a memoir of a life in computer science, so a homeopathic-level discussion of computer science is included. If any hint of tech talk gives you the hives, go ahead and skip this appendix, which is about software in VR. But you might surprise yourself and find it interesting.

A question: What should VR software be like? It ought to have an entirely different form than other software. Here's why.

Almost all software exists in two phases, like caterpillar and butterfly. In the first phase the software is written or tweaked, while in the other it is run. Programmers take code back and forth; tweak it again, run it again. The two-phase nature of software is practically universal. In a given moment a programmer either is writing a particular piece of software or observing it run.

(It is true that there are "builder" games like Minecraft in which you can change a lot while you're also playing, but there is typically a limit to how much change can occur before you have to switch to the caterpillar mode to make deeper changes.)

But this feels inadequate for VR. VR doesn't run in an external box, like your smartphone. You're in it. It is you.

Consider the physical world, say, your kitchen. When you first cook and then eat, the rules of reality don't have to change in between activities. You're not put into suspended animation while technicians come and reengineer your hands to operate a fork and knife instead of a skillet and spatula, or at least we have no reason to believe this happens. You simply do one thing and then another, within the same world, the same continuity. Wouldn't it make sense for VR software to be like that as well? Modeless?*

This was obvious from the start. So my compatriots and I had to reconsider the architecture of our software, starting with the most fundamental principles.

Grace

Back-and-forth mode swapping—between developing and running code—was mostly invented by Grace Hopper, the navy rear admiral and computer scientist who "codified" the core patterns for how software is still created today.

"Source code" is the artifact we modify while in the caterpillar mode, when computer software is being created and edited. Such code is typically made of words from English plus other symbols, and it seems somewhat readable, as if it were a story about what the computer is supposed to do. But that impression is deceptive. It is more like a legal document detailing the precise course of action the computer must take in order to not fail.

The stylistic misdirect often confuses students writing a first program. Even though it looks a little like human-friendly text, source code actually works only if you compose it with obsessive, robotic precision. You have to become a robot to program a robot.

The perfection of the idea of source code was mostly a result of the amazing work of Hopper and her all-female team of navy mathematicians, who invented or perfected programming languages, compilers, and the other technologies needed to implement "high-level" source code.†

The top male mathematicians had been corralled in Los Alamos, New Mexico, to figure out how to make an atomic bomb, so only the female mathematicians were left to further the cause of computing. Hopper's team

---

* It's an insider's reference to Larry Tesler's famous license plate in the 1980s, NO MODES. Modes make software harder to use. Larry invented browsers and many other familiar elements of our digital world.

† "High-level" has a drifting meaning, but usually means more distance from the bits; more engagement with abstractions that summarize what the bits do.

was spectacular, even creating an optimizing compiler, way ahead of when that would become a hot topic in computer science.

Text-based code demands that one particular abstraction become dominant, for it will provide the vocabulary. Therefore Hopper's approach had the effect of making abstractions seem fundamental and unavoidable.

Picture This

Most of the earliest computers, such as one churning in John von Neumann's basement lab at the Institute for Advanced Study at Princeton, included a rudimentary visual display: a light for each bit so you could watch it flipping moment to moment.* You could watch a program running.† That's how I like to think about computation, as a concrete process involving materials changing states; the flipping bits.

It's conceivable that a different way of programming computers might have come about if engineers had decided to try to make those lights more useful. Imagine this: The obscure, primordial visual array of bits flipping could have gotten better and better, to the point where you could paint and repaint the bits on a screen, so that a program could be redone as it was running.

How could this work? How would you know the meaning, or the implications of what you painted? How would you know what bit does what?

How would you keep the computer from crashing? How could your painting be perfect enough? Remember, even the slightest mistake can cause the computer to crash.

The bits couldn't just appear as a meaningless jumble. They'd have to be organized into meaningful pictures. There would have to be a remarkable (forgive the pun), highly constrained method of painting.

Please suspend disbelief for a moment about whether this would be practical, desirable, or even possible.

I suspect that if computer programming had evolved along these lines,

---

* http://alvyray.com/CreativeCommons/AlvyRaySmithDawnOfDigitalLight.htm

† The inside of a classical (non-quantum) computer is nothing but a very long list of switches that are either turned off or turned on. We call those switches bits, and speak of them as being in either a zero or a one state. While a computer is running, the switches are getting switched a lot. That's all that's going on in a computer. The rest is interpretation brought about when we perceive peripheral devices, such as when a bunch of the bits are shown on a screen as an image.

the whole society would be different today. The main reason might be a little hard to understand at first, but I'll come back to it: When you can see the bits and manipulate them, you get a more physical and down-to-earth feeling about a computer.

Source code, however, is not down to Earth. It is all about abstractions associated with a particular computer language. It makes us commit to those abstractions so consistently that denizens of digital culture start to believe in them, and perhaps become more vulnerable to believing a little too much in other abstract entities like AI beings or supposedly perfect ideologies.

Leaving that hypothesis aside, a more concrete, visual, and immediately editable style of computation would be *modeless* and better suited to VR. You'd be able to change the world while you're inside. Much more fun!

But what I have just described is only a fantasy of what might have been. The concept of source code programming took over.

There's a lot to like about source code. You nail down the state of the software each time you test it, so in theory at least, it's possible for testing to be more rigorous. (In practice, software remains hard to debug, but that's another topic. For those who don't know, the term "software bug" came from a moth that disrupted a program when it got caught in one of Hopper's early computers.)

I met Hopper a few times, and couldn't respect her work more— actually she intimidated me, to be honest—but here we have a great example of how computer science forgets that there are paths remaining to be explored. There was never a reason to think that all software always had to follow the pattern set by Hopper.

Sleight

The artificial divide between the programming and execution of code was a side effect of the very idea of text-based code. It's not intrinsic to computation.

Could the alternative history I described still come about in the future? Could there be a user experience that lets you reset the bits in the computer that comprise a program, *while* the program is running, without having to commit to immovable abstractions?

If that method became sophisticated enough, maybe programming could become more experimental and intuitive. That in turn would open up the path to reconceiving programming as a way of expressing whole worlds, systems, experiences; of expressing new levels of meaning that we couldn't yet articulate. That is what I wanted from computers.

My term for this ambition is *phenotropic*, though it's also sometimes called *neuromimetic* or *organic* programming. "Phenotropic" suggests surfaces turning toward each other.

Phenotropic software is still an experimental idea. There was a brief burst of experimentation in the earliest flowering of commercial VR. For instance, in VPL's virtual world software, the contents and the rules of a virtual world could be changed in any way, fundamentally, while you were in the world.

We accomplished this using a tricky mechanism that I won't attempt to explain here. It was a misdirection scheme, in which we could swap out new bit patterns for old bit patterns just in the moment when the central processor wasn't looking at them. The trick had to be executed perfectly, since a lot of bits have to change in just the right way at just the right moments so as not to crash a machine. (Everything has to be perfect at the bit level for computers not to crash.)

We engaged in these heroics initially because it was the only way to get fast-enough performance from the slowpoke computers that were around back then. It happened that we could often make code run faster using the mechanism.

It was only a pleasant side effect, at first, that we could also change what a virtual world did while we were inside it.

Editor and Mapping

The components that make up a phenotropic architecture are called *editors*. It can be a little hard for computer scientists who are used to conventional architectures to get comfortable with the concept at first.

The main difference between the experience of phenotropic programming and the current, familiar variety is that a phenotropic programmer doesn't have to look at the same format of source code over and over again.

Currently, all code of a given programming language looks similar. Endless occurrences of IF, THEN, REPEAT, or whatever words and symbols are specified.

In a phenotropic system, there are different, specialized user experiences for different aspects of a program, and for different kinds of programs.

These designs that you perceive and manipulate while programming phenotropically are called *editors*. An editor might look like images on a computer screen, or virtual objects in a virtual world.

An editor is a *mapping* between a user interface experience and patterns of bits.

If you're editing the bits of a program while it's running, that means the *editor* you are using has to be able to interpret and present the bits to you so you can understand how to change them. There might be more than one way to do that. Different editors can be pointed at the *same* bit pattern, the same program, and present it to a programmer in different ways.

Since phenotropic programming is based on mapping between the human experience level and the bit level, programmers are less committed to particular abstractions. One editor might map a bunch of bits that run as a program in such a way that they look like a maze, while another editor might map the same bits so that they look like a family tree.

Each conventional, source-code-based programming language is inexorably tied to its abstractions, like Fortran's functions, LISP's lists, or Smalltalk's objects. These are all examples from the era when I learned to program. You don't need to know what they were. The point is that they were concepts that bridged the world of human intent with the flipping bits inside a computer, and each one shone in certain circumstances and became awkward in others.

Phenotropics is an approach that supports different such concepts at different times within one tool. Abstractions can be mixed and matched to suit the needs of the moment.

Variation

This doesn't mean that abstractions become obsolete.

Imagine living in a future in which VR is programmed in the way my friends and I explored so long ago, phenotropically. In that case, there will be various actions you can perform that cause the underlying bits to change so that the virtual world you are in will suddenly function differently.

What might these actions be? Will you operate a simulated control panel that looks like the bridge of the Starship *Enterprise*? Or will you pull on chains in a medieval dungeon, or dance like a leaf? Or edit text that looks like the Grace Hopper–style source code everyone uses these days? Any and all of these editor designs might have a place.

But there has to be *some* design. You can't accomplish anything without embracing a point of view and a way of thinking. But there is no reason, fundamentally, to be inflexible about which design to use at a given time.

What has happened in Grace's universe of nonphenotropic source code is that each computer language proposes that certain abstract objects are not only real but mandatory, eternal, and inevitable while using that language. I already mentioned the classic "functions" of Fortran and "objects" of Smalltalk, but I could just as easily add the "bots" in cloud software that are fashionable as I'm writing this book.

Each of these objects is fine and useful at times, but none of them ought to be inevitable. They aren't real, if "reality" means that which you cannot reject. I find it troublesome that they *seem* real.

Different abstractions might have been put in place instead of any of the familiar ones, but for the quirks of history. (Whether widely used software abstractions can ever be reconsidered is an open question. In a previous book, *You Are Not a Gadget*, I examine the way ideas that are expressed in software can become "locked in" by pernicious "network effects," but for the purposes of this book, I assume that there's time and hope for change.)

The only things that are fundamental and inviolate—truly real—while you are using a computer are you and the run of patterns of bits inside the computers. The abstractions linking those two real phenomena are not real.

Is it imaginable that a computer architecture could express this philosophy? What if there was a way to swap in and out different editor designs in order to present given bit patterns to you, so that you could understand them and modify them in different ways at different times?

Phenotropic Trial Run

My friends and I built a few generations of phenotropic experiments in the early 1980s. An early one was called Mandala, then came Grasp, then Embrace. (Grasp suggested a glove, while Embrace embraced whole body suits, literally and figuratively.) Some of the major categories of VR applications were first prototyped in VPL's codeless software.

"Codeless" was not a metaphor; in a literal sense, we didn't use code. Well, we used traditional code and development tools to get the system going in the first place, but virtual worlds didn't run on code, only on bit patterns that could, to repeat, be modified by *editors* that *mapped* to them.

Editors are fundamentally different from the usual tools used to create software, like compilers and interpreters.

Compilers are the chrysalises in the metamorphosis scheme of traditional code-based software: You edit a text file, the "source code," and then, only after compiling, you get to see what the code does once it has been changed. Then you go back and forth to debug.*

The phenotropic alternative might sound like an impossibly exotic idea to younger computer scientists who have grown up in Hopper's shadow. The idea of code is almost universally taken to be synonymous with computing, but it need not be.

Could a phenotropic editor mimic traditional code? In other words, could we edit bit patterns by mapping them to images on a screen that looked like a familiar high-level text-based language? In many cases we could, which meant we *simulated* code. A phenotropic editor could be designed and constrained to look like text, even though the effect arose from a more general graphical construction. Such an editor could do anything a compiler could do, but as live visual tweaking.†

We played favorites with certain editor designs, meaning with certain visual representations of code, often preferring a principle called dataflow. Dataflow typically looks like wires connecting modules. But it wasn't fundamental. We could swap in Grace Hopper text–like editors, or other editors.

The experience of programming briefly became a little more improvisatory, a little more like a cross between playing jazz on a horn and drawing mathematical diagrams.

---

* Interpreters are similar to compilers in that they also are keyed to a fixed vocabulary and grammar of text-based code; a particular computer language. Instead of directly running on the bits of the computer, however, interpreters run on a simulated computer that is itself a program running on the real computer. That means they can allow for changes to the program while it's running, since the real computer isn't running the code (so there's no worry that the real computer will crash). The disadvantages of interpreters are that they can be slow, because of the layer of indirection, but more important: As with compilers, there's no way to change the abstractions in play; the abstractions are fixed in the design of the language.

† Often we could make small changes so quickly that the experience of change was in real time, though overhead varied per tweak.

Fifty-second VR Definition: A way of using computers
that suggests a rejection of the idea of code.

Alas, we eventually had to ask VR customers to develop on a regular monitor instead of from within the virtual world. The main reason was that regular monitors were so much cheaper than VR headsets. More people could be working at once and in more locations.

It still hurts today when I think about that! It hurts worse that everybody in today's VR renaissance is still using conventional programming languages on conventional screens to develop for VR. It's like trying to learn a foreign language from a book without ever talking to a native.

Our editor designs on traditional monitors often looked a little like MAX, a visual programming tool used today for experimental computer music and animation.*

We at least peered into an alternate future that will hopefully be explored more thoroughly in years to come.

Scale

A fundamental impulse in computer science is "scaling." That means computer scientists hope our works will be able to scale up to become unboundedly huge and complicated.

How do you make bigger and bigger phenotropic structures? A phenotropic editor maps the machine's bits to a user interface so a person can change the bits. But can one editor edit another editor? Could you have towers of editors editing editors, webs of them, giant fungal growths?

Yes, of course, that's the idea. But in that case, would you need to commit to a set of abstract principles each editor would adhere to so that it

---

* MAX looks like a scattering of little boxes with a web of lines connecting them. It roughly mimics the experience of programming old-fashioned synthesizers like the ones Bob Moog and Don Buchla designed. (Those were operated by plugging a lot of cables into jacks in metal boxes mounted in frames.)

    It was named in honor of Max Mathews, who invented digital audio back when he was at Bell Labs. While Max was alive, we used to have breakfast every Thursday in Berkeley that also included Don Buchla, Tom Oberheim, Roger Linn, Keith McMillen, David Wessel, and a rotating gang of other pioneers of electronic music products. This was the crew, along with Bob Moog, that had been my model for starting the VR industry.

could be edited by other editors? Wouldn't that defy the goal of avoiding commitment to any particular abstraction?

The answer, incredibly, is no! A phenotropic editor need not adhere to any particular abstraction to be operated by other editors.

The reason why is that each editor is a user interface, usable by humans. Therefore, editors can act like simulated humans to operate other editors. An editor can interpret a user interface and use it on the terms of that interface.

For instance, an editor for low-level access to a math library might look like a calculator. A person could use it directly, or another editor could use it through a simulation of user interaction.

A calendar program that needs to call upon arithmetic to calculate the date for a future appointment would simulate pushing the buttons on the simulated calculator.

There need be no shared abstraction dictating how one program is supposed to call another. Instead, each editor becomes responsible for figuring out how to use the human-oriented user interfaces on other editors.*

It might sound like an uncertain and highly inefficient way to get one part of a program to interact with another part, and it is! But only for small programs.

The phenotropic hypothesis posits that once you start dealing with very large systems, with giant programs, a phenotropic principle becomes *more* efficient than a traditional one, which must enforce abstractions.

You can think of a phenotropic system as a bunch of editors with simulated people sticking out the back of each editor. In a couple of our old designs, you could turn a whole big program sideways to see the collection of underlying editors edge-on, floating in a formation in space like shields in a space war.

Emerging behind each editor was a cartoon-like character in profile, animated as it operated other editors that in turn were backed by other characters. This was all done in the era's only practical style—8-bit game graphics. We never quite got the whole vision implemented, but we got close. I wish I had an image to show you, but apparently none have survived.

---

* The engineering is similar to efforts to create programs that can probe and use devices without prior instructions. An example of this related work is found here: https://cacm.acm.org/magazines/2017/2/212445-model-learning/fulltext

That sideways view was just another editor, of course. Nothing special.

(If you've already read the section about my take on artificial intelligence, consider this: Unlike AI, in which a simulated character faces you, in a phenotropic system they all face away from you, toward other editors, but under your control. It is clear they are tools, not equals to you. Same algorithms as in AI, but different conception.)

Motivation

There are varied reasons to entertain the phenotropic hypothesis. Before getting to the nuts and bolts of efficacy, consider usability by people.

It's always easier to write a fresh program than to understand and modify someone else's program, but at least, if the program is phenotropic, the pieces you find when you open the hood are always user interfaces designed for people. Because that's all there is.

A phenotropic system will tend to be made of components that are the right size for human use, since each editor is initially designed for human use. That means phenotropic systems tend to have "coarser chunking" than other architectures.

Instead of a zillion little abstract functions, the organization of a large program will be broken into larger, lucid pieces, each of which is coherent on its own as a user interface. The chunking naturally follows practical human usability instead of an engineer's idealized scheme, and will tend to be easier to understand and maintain.

In a phenotropic system, it should be possible to watch the animated characters behind each editor doing what they do to get a sense of how an overall program works, but you can also position yourself anywhere in the network of editors to directly play within the program, to experiment.

This observation suggests a fundamental principle. Computers only make sense as tools to serve people. If you make a computer "efficient" but that efficiency makes it harder for people to understand and maintain sensibly, then that computer actually becomes inefficient.

Role Reversal

A great example of this principle is found in computer security. We have created endless layers of abstraction to allow programs to communicate with programs, but those abstractions are hard to understand. Therefore, hackers keep on coming up with unforeseen exploits, and we must all accept an astounding overhead in breaches, maintenance, security software, election tampering, identity theft, blackmail, and on and on.

Would phenotropic software really be more secure? I can't prove it unless and until there are more tests, but I'm optimistic.

Here is how we build systems today: A bit-precise structure of communication abstractions surrounds the "pay dirt" modules like "deep learning"* ones that accomplish the most valuable functions.

These critical "AI-like" algorithms are *not* bit-perfect, but even though they're approximate, they're still *robust*. They provide the capabilities at the core of the programs that run our lives these days. They analyze the results of medical trials and operate self-driving vehicles.

In a phenotropic architecture, the roles of the bit-perfect and approximate/robust components of a program are often reversed.

Modules are connected in a phenotropic system by approximate but robust methods like deep learning and other ideas usually associated with "artificial intelligence."

Meanwhile, bit-perfect precision is invoked only inside certain phenotropic editors, like the functions accessed by the calculator. Absolute precision is no longer used for communication.

Why would this be more secure? To protect a computer from hackers, we sometimes create an "air gap." That means that a computer performing a critical function isn't even online. It's out of the reach of hackers. A real person has to use it on-site.

Each and every module/editor within a codeless phenotropic network is surrounded by what is effectively an air gap, because they can't receive abstract messages from one another. There are no messages. Only simulated fingertips pressing simulated buttons. There's no abstract "press button" message.

---

* The terminology for algorithms that can interpret images and other natural data is in flux. During the period I recall in this book, the term "pattern recognition" was prominent, while in the new century, "machine learning" became more popular, since it was associated with some more effective ideas based on access to larger data sets. More recently, "deep learning" is ascendant, and it is associated with another step in making more effective algorithms. The terminology will keep on shifting as scientists attempt to distinguish their advances from those in previous generations of algorithms. These differences in algorithms and in the terms associated with them are not critical to the phenotropic argument, so I am using them loosely.

Before I get back to the topic of security, I'll explain more about how the air gap works.

Expression

First, a confession: Back in the eighties there was no way to implement the phenotropic effect without a "press button" event. Machine vision and machine learning were not working well enough yet.

So we needed a little language to describe display and user interface features like onscreen buttons, but we knew that it was only a temporary patch for a temporary problem.* Moore's Law suggested that eventually computers would get fast enough to be able to recognize *similarity* instead of only *identity*. When that happened, then one editor would be able to watch another with machine vision and operate it with a virtual hand, and there would no longer be a need for an abstract representation of a user interface element like a button.

In the mid-nineties, when computers finally got fast enough to recognize visual similarities in real time, a new batch of friends and I founded a startup, called Eyematic, to perform machine vision tasks like recognizing faces or tracking facial features. (We won government contests run by the U.S. agency NIST in that era for recognizing and tracking faces in difficult real-world situations.)

Most of the scientists on the Eyematic team were ex-students of the formative neuroscientist Christoph von der Malsburg. A few of the original Veeple also reconvened there, including Chuck and some of the same old investors, though the heart of that company was Hartmut Neven. Google bought it eventually.

I have to admit that it was disturbing to work on some of the first effective facial tracking and recognition programs. Were we creating a monster? I used some of the Eyematic prototypes to implement working models of nasty technologies that became scenes in *Minority Report*, like

---

* 1980s-vintage phenotropic experiments relied on a scheme that described everything you could see in a virtual world or on a screen, from rooms, to avatars, to text, to windows, to icons. It consisted of five primitives, each of which described a visual/spatial relationship such as containment or order. Using this system, we described and rendered everything, including even the simulation of traditional source code, though under the hood, there wasn't any source code. Instead there was just a particular mapping from machine language (the way bits were set for a program run) to what was on the screen.

the billboards that incorporate someone who is running past them as he attempts to evade the police, broadcasting his position to everyone.

The reason I persisted is that I felt there was a benefit that would more than make up for the ugly potential for universal surveillance. If we could get machine vision to recognize faces and track expressions and so on, could we not also apply the same abilities to get editors to use other editors? We could finally get rid of the temporary patch and build a proper phenotropic system, with proper air gaps.

In that case, a phenotropic editor would not support any interface or method of interaction other than its user interface. There would be no protocols, no abstract variables to document, no API.*

Machine vision and machine learning algorithms from one editor would be applied to interpret and operate a virtual hand that would virtually touch another editor. An editor could not "tell" if it was being operated by a person or another editor at a given time, because the interface would be identical in either case.

The nature of the code within an editor that causes it to do certain things to another editor would not be standardized. Nor would the means of programming a given editor.

Some editors might be trained to perform (in the way that we train machine learning algorithms with examples) while others might have to be explicitly programmed. All could interface with all just as a person could.

My belief is that this prize is big enough to outweigh the problems of surveillance. If our information systems can be built on principles similar to the phenotropic ones I have described, then we can eventually come to use tools that don't require us to accept only certain abstractions universally and forever.

Since from now on our information systems will serve as the molds for many aspects of society—and the guides for how young people become individuals within a society—a move to an information architecture with varying, revocable abstractions couldn't be more significant. This is how openness and freedom might be encouraged in the far future.

I realize this hope might sound esoteric, and might also sound like a

---

* An API (Application Programming Interface) is a typical current way to package abstract layers used to connect programs to one another.

humongous leap of faith, even a utopian impulse, but it's actually an attempt to outgrow utopia.

Leaving aside the big ideas, it was great fun to finally get avatar faces to track the expressions of human faces. For a while, I experimented with expressive avatar faces with my weird band in clubs like the Knitting Factory in New York City in the 1990s. There was a big screen behind us, and kooky musicians were seen morphing into avatars of the corrupt politicians of the day, for instance. (In retrospect, corruption was pretty tame in those days.)

The Wisdom of Imperfection

Since the modules of an ideal future phenotropic system would be connected through approximate means, using machine vision and other techniques usually associated with artificial intelligence, a lot of the manic, tricky hacking games that go on today wouldn't even get off the ground.

It would be hard, for instance, to inject malware into a computer through a deep learning network, say, by pointing a camera at an image that is supposed to cause the infection. *Hard* is not the same thing as *impossible*, but the quest for perfection in security is a fool's game.

To be clear, you can inject malware using an image (it's done all the time), but it's only easy to do that when software ingests the image bit by bit and processes it using a precise protocol.

It's easy to fool lockstep protocols, because you can usually come up with a trick the original designers didn't foresee. A common example is that more bits are placed in an image than are identified in the protocol that delineates the image. When the image is ingested, some of the bits overflow into a part of the computer where they weren't expected, and those extra bits can contain malware.

The infection of computers using strategies with this flavor is possibly the most common human-caused event taking place on planet Earth at this time.

But if an image is ingested *only* as an analog-style approximation and only analyzed statistically, as if a camera had been pointed at it, then there is much less vulnerability.* The image is not the problem; the rigidity of the protocol is the problem.

---

* One of the prototypes of a phenotropic system included actual small screens and cameras hidden in a machine, so that the air gap was implemented physically.

Sometimes it's best when engineers don't know *exactly* how software works.

The approximate nature of modern algorithms associated with "deep learning" and related terms is inherently resistant to the tricks of the hacker trade, but we apply those capabilities only to performing specialized tasks, not for building architectures. Another way of framing the phenotropic idea is that we should use them in architecture.

Just like in biology, security is enhanced when a system becomes robust, which is not the same thing as perfect. A perfect system shatters, while a robust system bends.

Resilience

Part of the phenotropic hypothesis is that systems that use AI-like algorithms for structure and connections instead of only for payload will be less vulnerable to constant catastrophic failures.

Of *course* phenotropics is a whopping inefficient way to link modules in a small system. You're relying on machine vision and learning algorithms to unite the most basic tasks. But maintaining protocols becomes inefficient in very large systems. There are constant updates and sweeps for viruses, for instance, and a lot of down time whenever a protocol needs to change.

An example I like to bring up comes from music. I have bought thousands of dollars of software plug-ins for music over the years, for tasks like adding reverberation to a mix, but none of them still work. *None* of them!

Software components become obsolete quickly because they depend on perfect conformance to protocols and other aspects of a software ecosystem, and it's hard to prevent the appearance of tiny changes.*

Meanwhile, I have also purchased a multitude of physical music effects pedals going back to the 1970s. I also have an extravagant number of physical music synthesizer modules. Many of these hardware artifacts contain computer chips that perform exactly the same functions as software plug-ins I have bought. But they aren't equivalent in one crucial way. All of the physical devices still work. *All* of them.

---

* In my case, the problem is creeping changes in Apple's Mac OS.

The difference is that the physical devices have analog, air gap connections that are resistant to obsolescence.

In theory, the software plug-ins should be cheaper, more efficient, better in every sense. In practice, the hardware boxes are cheaper, more efficient, better in every sense, *because they still work*. Hardware effects pedals and modules are the phenotropic version of music technology,* while plug-ins are the protocol version.

You can't just look at the way a technology performs in one moment of time. You have to look at the whole life cycle, including development and maintenance.

My experience with music tools illustrates another aspect of the phenotropic hypothesis: A phenotropic architecture will become more efficient than a protocol-centric, code-based, traditional one at a sufficiently large scale, and when considered over a long period of use and modification.

Adapt

The word I always used to use to criticize conventional computer architectures was "brittle." They break before they bend, even if just one bit is wrong.

For an alternative to brittleness, look to life. Consider how natural evolution is able to function. Our genes can be a little like software sometimes: Occasionally a single mutation can be deadly.

But it's completely normal for different individuals to be viable even though we don't have identical genes. Small changes don't always crash us.

We don't understand genes entirely, to say the least, but this much is clear: They are robust enough to allow evolution to happen.

Evolution is an incremental process. Small changes accumulate over deep time to turn into almost unfathomable changes. From single-celled organisms to us.

The crucial microstep along the way is when a small change in genes results in only a small change to the resulting organism. This correlation of small to small happens often enough that the feedback loop at the core of evolution has a chance to function.

---

* To avoid any potential for confusion: It's not the fact that pedals are hardware that makes the difference. It's that they connect without having to adhere perfectly to protocols and other aspects of a digital ecosystem. The same benefit could be had in phenotropic software.

If small genetic changes caused organisms to change radically too often, then they wouldn't "teach" evolution much, because the results would be too random. But because the results of small genetic changes are also small in many cases, a population can "experiment" with a cluster of similar new traits gradually, and evolve.

Meanwhile, if you randomly change a bit you might completely crash a computer; if you *cleverly* change a bit you might compromise a non-phenotropic computer's security.

However: It's virtually impossible to flip a bit in an unforeseen way in a present-day program and create a small improvement. Doesn't that mean we're using bits wrong?

So: Another aspect of the phenotropic hypothesis is that small changes made in a phenotropic editor should result in small changes in its behavior often enough to facilitate adaptive improvement at a *large* scale.* That doesn't happen with current systems.

Swing

When I imagine future phenotropic systems, I imagine them to be spread out over a network, with editors operating each other across the globe. A cloud of cartoon characters poking at each other.

The operating controls for physical devices like thermostats and drones will be identical for cloud algorithms and people; therefore, there will be less esoterica to disable people from understanding gadgets.

I imagine beginning students browsing through the cloud architecture of the world, playing, tweaking, exploring, everything designed for people

---

* Should large phenotropic systems ever be built, a typical one will probably include a lot of redundant parallel paths of similar, but divergent, editors. (Not unlike how we authenticate people today, with multiple factors such as calling their phones in addition to having them enter passwords.)

Once two or more paths of editors have yielded comparable intermediate results, then specialized editors will compare them. Redundancy will compensate for the uncertainty of the imprecise statistical connections.

Redundancy will do more than improve reliability. It will also allow large systemic adaptation of an architecture, not just an algorithm.

Through redundancy, editors and collections of editors will be tested against one another, and the overall system will improve. If one path is working better, it will be favored, and might influence the design of new paths.

This mechanism will recall the value of genetic diversity within populations of living organisms.

Engineers evolve algorithms in this way already, but *not* the connective, architectural structures between algorithms. That omission will be corrected by phenotropic systems.

from the start, everything comprehensible. Of course, I imagine it all happening from within VR.

If people could change the way the virtual world worked while they were inside, then . . . well, here I still gasp and search for words. This was the state I did my best to describe in my ancient talk on "postsymbolic communication."

I used to say that a mature twenty-first-century VR would be a fusion of the three great arts of the twentieth century: cinema, programming, and jazz. Of these, the jazz element will be the most challenging.

Jazz is improvised. Musicians make it up spontaneously.

We have seen preliminary tools for rapid creation of computer content. People use their smartphones all the time to compose contemporaneous expressions made of text, photos, movies, and sound recording. It's amazing how fast the more agile users, especially kids, can modify virtual worlds in builder games like Minecraft. But deeper improvisatory programming remains an elusive idea.

In principle, a "deep" or convolution network similar to those that can translate languages or interpret images might be able to modify programs adaptively, so that a user could guide a program through changes by dancing or playing a saxophone.

You'd then be able to define a new interactivity or physics for a virtual world with the kind of agility and speed we're used to today when we speak a sentence or make a dance move. Forming the words to be spoken or planning a dance move does take time, but they happen about as fast as we think and feel, so they feel "real time."

Could programming ever be real time? I've played around a lot with ways to help people quickly "feel out" programs as a means of creation instead of having to compose them line by line.

One way is to let people be selective instead of constructive. The easiest way to understand this is to consider a purely sonic version of the idea. Suppose you bombard a user with a cacophony. Whenever the person moves a hand, the sound occurring at that moment starts to recur as part of a loop, and other sounds that weren't chosen become just a little quieter. Then the person repeats the process, selecting more sounds that join the repeating loop while allowing other sounds to fade away.

At the end of the process, the person, who might be a nonmusician, will have created a loop of sound that is genuinely their own composition, and will vary more, and be more personal, than the usual novice compositions that come out of programs like Garage Band. But there was never a process of invention, only selection, though one can argue all night about the distinction.

The approach can also be realized in visual design, where a person might pinch on little swirls that go by within a noisy, windy virtual world, solidifying them, and continuing to do so until a sculpture remains. It would be a little like an active version of the old Rorschach test; but an experience that starts from scratch.

So can a strategy along these lines be applied to general programming? So far, I've never achieved a convincing demonstration, but I still hope it can be done. The main problem is that it takes time to observe a behavior and programs describe behaviors.

I've long looked to musical instruments for inspiration in user interface design. If you think of technology as a form of expression, then there is no question that musical instruments are the most advanced technologies that have ever existed, although this compliment applies only to non-digital instruments thus far.

It always astonishes me that people can improvise jazz. Improvising involves problem solving of considerable depth that goes on in real time. The programming of the future will have to be a lot like jazz.

That is why I've put so much effort into contraptions like virtual saxophones. They don't play nearly as well as real saxes, at least not yet, and certainly not as well as really fine saxes.

Some of the problems that I imagine future phenotropic improvisers will have to overcome are different from the ones facing a modern-day jazz saxophonist. If you have to watch a million behavioral variations, perhaps of how a flying alien turtle flutters around your head, it would take years to observe enough to choose the best.

Suppose you saw hundreds, even thousands of instances of the turtle, each transparent, forming a turtle behavior cloud, in which you are to select the behaviors within the storm that seem most salient. Maybe an instrument, a virtual sax, let's say, could be played to sync with the motions of the turtles you want to highlight.

From a phenotropic perspective, we need to learn how to design "editors" that convey the range of what a program can do in a convenient, viable manner. People will spend centuries on the problem, mark my words.*

The established way we cope with very large numbers of concrete possibilities is through abstraction. The open question is whether there's a more fluid form of concrete expression that can become a practical alternative to abstraction. It's only imaginable in user interfaces beyond what we know, probably in future versions of VR.

Rubble Fills Plato's Cave

Computer science can be thought of as a branch of engineering, or an art, a craft, or even a science. It's all these things, but mostly, to me, it's applied philosophy, or better yet, experimental philosophy.

Computer scientists have ideas about the meaning of life, and the practices that make life good, and implement those as patterns that will guide the real lives of real people. Usually, we are idealists. Since ideals are never fully realized, the history of computer science can be understood as a boulevard of broken dreams.

I know full well that phenotropics might never get a full hearing, that it might be too late. Even if there's a massive revival of phenotropic research, there will turn out to be problems I never foresaw. Nothing is ever perfect.

But this is the mind-set of computer science. You keep chasing. Ivan Sutherland has been pursuing "asynchronous" computer architectures for years now. These are hardware systems without a master clock, but the implication is deeper; that computation can be fundamentally less localized and hierarchical. It's been a long haul for him. Similarly, Ted Nelson is *still* working with a shifting group of students and followers to implement Xanadu, the original design for a digital network, which he started on in 1960. I'm convinced it would be better than the World Wide Web, but no one can know until there's a fuller implementation.

The idealist projects of computer scientists aren't the ones that end up running the world, but they have indirect influence. Bits and pieces end

---

* It will never happen for all programs, but only for programs constrained to have a representable range of behaviors.

up inserted in odd spots. The World Wide Web is a pale shadow of what Ted originally proposed, but it is informed by his ideas.

While math is an ever-rising tower of truths, computer science is more like a mound of fragments from forgotten wars. This is not a lament. Another difference is that computer science practically mints money, and it will continue to do so until computer scientists make money obsolete.

A sad side effect of the wealth, though, is that it focuses so much attention on whatever the top companies are doing at the moment, forsaking other ideas. The rich variety of computer science idealisms don't get as much of a hearing in the wider world as they deserve. Each one represents an alternative reality we could have lived in, or might live in one day.

Many Caves, Many Shadows, but Only Your Eyes

Having experience with simple phenotropic experiments changed my philosophical outlook. For one thing, I realized how wrong it is to treat software as real. Bits are real, since they're measurable in chips or as they're transmitted, and people are real, but the bits have meaning only because of people.

I used to say "Information is alienated experience." That is, bits have a meaning based on human experience when they are put into a computer, or when they are extracted, but absent human culture and interpretation, they're meaningless.

Another way to state the case is that to an alien, there's no difference between a smartphone and a lava lamp. Each gets hot as patterns evolve in it.

I find support for this point of view in the Fermi Paradox. How is it possible that we don't see evidence of other life in the universe when we sift through the night sky? Maybe because life is hard to recognize if you don't share a culture. We see noise where aliens might see literature.*

Once you see that bits never take on intrinsic meaning, it becomes easier to make computers better, because then the only criterion left is to design for people.

This way of thinking elevates the status of people. There must be something special about us. I can live with that.

---

* A clarification: I see encryption as a form of culture. A way to interpret bits. So aliens with encrypted signals would be undetectable, but the difference between encryption and a sufficiently alien foreign language is moot.

Appendix Three:
Dueling Demigods

Aldous Huxley's *Brave New World* anticipated the dark side of VR with a fictional media technology called the "Feelies." I didn't get around to discussing Huxley's vision because one can't put everything in a single book. It's a special shame in this case because Ellery lived with Huxley for a while, in Southern California, and Huxley is also famously relevant to the culture of psychedelic drugs. Once I thought this book was completed, Donald Trump was elected. I felt compelled to write the following text, which echoes Huxley, but with a level of detail specific to the times. I have no idea how it will read once the book is published, but I include it here to capture the moment.

Not Artificial, but Imaginary

Remember how, in the chapter called "How We Settled into a Seed for the Future," I wrote that even if robots and algorithms take away all our jobs someday, they still won't really be doing anything? All the information in a robot or a cloud algorithm ultimately comes from people. All the value. You are being picked apart for data all the time, and that data will be used to drive multitudes of machine learning schemes that will put you out of work.

The easiest-to-explain example of this principle relates to automatic language translation, so I use it all the time. The Internet has decimated the livelihoods of professional language translators, just as happened to recording musicians, investigative journalists, and photographers.

But if you look carefully at how automatic translation works, you will learn that the algorithms have to gather the real life translations of real live people by the *millions* every single day to serve as the example sets. (Public events and pop culture move on a daily basis; so does the language.) The algorithms *seem* to be self-sufficient, but they are actually repackaging value that comes from hidden individuals. Pull back the curtain of AI and there are millions of exploited people.

I am not saying AI is bad! I'm saying it's not a thing at all. Fearing AI is just another way of amplifying the harm done in the name of AI. It's just as fantasy-bound to fear a mere algorithm that people want to use—to worry, for instance, that it will *inevitably* create unemployment or a crisis of meaning—as it is to pretend that the algorithm is alive and would be valuable on its own without the stolen human data. The only way to reduce the harm is to stop believing in AI as a new type of creature, but instead evaluate algorithms as tools to be used by people.

Automatic translation services are useful. It would be counterproductive to fear them and shut them down. What would be ideal, ethical, and most important, sustainable, would be to celebrate and pay the people who are providing the data—in this case, translated phrases—that make the algorithms possible.

In the 1980s, many of my friends *liked* the idea of a future of pretend economic valuelessness; then everyone would be forced to accept either a pure form of socialism or some other utopian scheme. More recently, this sort of thinking has reemerged in discussions of the Basic Income Model (BIM), which would grant all the useless people stipends once the robots start doing all the work.

(Even more recently, in the age of the "alt-right," I see the same strategy being plotted in some of the harsher hacker circles, but this time to get people to accept some sort of racist autocracy. When the robots put everyone out of work, goes this line of thinking, then ordinary people will have nowhere else to turn.)

I suspect that the Basic Income Model is a trap. People would feel useless and absurd, economics becomes destructive when value is selectively ignored, an ersatz social safety net that can be gamed by speculators is not a safety net at all, and a centralized, superpowered political body would have to command the scheme; an invitation to corruption.

I've written about those concerns in *Who Owns the Future?* Here's a lightning summary of my argument: AI algorithms depend to varying degrees on frequently replenished big data, but overall the project of AI is dependent on stealthy access to mass human data without acknowledgment, much less payment. *WOTF?* proposes that bringing the hidden data value of people into the formal economy through nanopayments—as an alternative to the current stealth/barter Internet economy*—can provide an alternative to the Basic Income Model. The motivation goes beyond concerns that the BIM risks devolution into a command economy. A universal data economy is an alternative to the BIM that will not only resist an unsustainable concentration of political power, but also reinforce individual creativity and human dignity.

Intuition might suggest that there wouldn't be enough income for each person, but remember that in the future there will be multitudes of diverse so-called AI algorithms generating nanopayments at once. Consider an extreme scenario in which all activity occurs through AI algorithms—meaning nothing is done directly by people. The value generated by human data would be at least as great as the historical value of people performing tasks directly. If there was enough value—and diversity of value—from people before, then there also will be in the future, so long as AI is interpreted as a repackaging of human capital rather than as an alien source of capital.

For example, consider a future in which teeth still have to be brushed, because genetic engineering, nanotechnology, or whatever didn't work out to make teeth brushing obsolete. But it's the future, so you don't brush your own teeth anymore; a robot interpolates the brushing practices of thousands of spied-upon people in order to brush your teeth to perfection. If we continue to inform robotic algorithms through stolen data, you'll sit there while your teeth are cleaned, feeling useless, kept, and absurd. However, if you know that a few special people out there are geniuses at teeth

---

* Even though wealth has only risen for elites in the United States during the era of big Internet computers running everything, there was also a dramatic reduction in abject poverty in the developing world during the same period. While I wouldn't suggest that digital tech explains everything, that good seems linked to cloud-connected gadgets, in the form of cheap mobile phones. Here's a hypothesis to explain why ordinary people did better in one case than in the other: People who use low-end phones for primitive texting and calls are first-class participants in a market, meaning they seek opportunity from individual perspectives instead of as directed by central algorithms. Another example was the way personal computers made small businesses more profitable during the era before the rise of the giant Internet companies. Now the same users are generally experiencing stagnation. Our algorithmic era is resurrecting command economy fallacies.

brushing and provided the examples that are making your teeth look and feel great now, and that those people are being paid, just like you are for what you're good at, and that they will be able to afford to pay *you* for what you do in a digital economy that is growing because people are creative . . . in that world, you won't feel useless or absurd. There should be dignity in everything, even the simple things like brushing teeth.

This way of thinking might elicit religious objections from true AI believers, but perhaps speculative metaphysics can be deferred for the sake of sustainable economics.

The Banality of Weightlessness

In "How We Settled into a Seed for the Future," I also described the formative imperative to make the experience of the Internet seem "weightless." One of the consequences, first emerging in alt. Usenet groups, was an explosion of cruel nonsense, because nothing could be earned other than attention, and no one had a stake in being civil.

Today, one of the biggest problems for virtual reality is that the immediately obvious customer base willing to spend money is gamers, and gaming culture has been going through misogynist convulsions.

This phenomenon is known as Gamergate. Complaints about how women are portrayed in games are drowned out by blithering barrages of hate speech. When a feminist game design is promoted, the response is bomb threats and personal harassment. Women who dare to participate in gaming culture take real risks, unless they can adopt a persona that puts men first. Gamergate has left a trail of ruined lives. And yet, needless to say, the perpetrators feel they are the victims.

The designs and culture emanating from the tech world don't explain everything, but they do have an immense influence.

For years, Gamergate was only a plague within digital culture, but by 2016 its legacy was influencing elections, particularly the one in the United States. Gamergate turned out to be a prototype, rehearsal, and launching pad for the alt-right.*

The kinds of problems that used to inflame only obscure reaches of

---

* https://www.theguardian.com/technology/2016/dec/01/gamergate-alt-right-hate-trump. It's vital to remember that tech is not the only force of our times (there's also tribalism, for instance), but it is the most commonly optimistic one, so its character has an exaggerated impact.

Usenet now torment everyone. For instance, everyone, including the president, is upset about "fake news." Even the news of the term itself was quickly made fake; the term "fake news" was deliberately overused to the point that its meaning was reversed within only a few months of its appearance.* It became the way a grumpy American administration referred to real news.

Fortunately, more precise terms are available. For instance, it's been reported† that the founder of the virtual reality company purchased by a social media company for a couple of billion dollars—that was mentioned in this book's opening—called the planting of sadistic online confabulations engineered to go viral "shitposting" and "meme magic." It was further reported that he spent serious lucre incentivizing the activity during the 2016 election. When you're paying actual money for something, you need unambiguous vocabulary to describe it.‡

Shitposting is clearly distinct from low-quality journalism or dumb opinion. Shitposting is one of those rare forms of speech that shuts down speech instead of increasing the amount of speech. It's like playing loud, annoying music in the cell of a captured enemy combatant until he breaks. It clogs conversations and minds so that both truth and considered opinion become irrelevant.

There have been widespread calls—from the entire political spectrum—for the tech companies to *do* something about the prevalence of shitposting. Google acted first, and despite initial reluctance, Facebook followed. The companies now attempt to flag shitposts, and they refuse to pay the sources. It's worth trying, but I wonder whether this approach addresses the core issues.

Consider how odd it is that the whole society, not just in our nation but globally, has to beg a few tightly controlled corporations to allow usable space for sincere news reporting. Isn't there something strange, perilous, and unsustainable about that, even if those corporations are enlightened and respond positively for now?

---

* The mainstreaming of Notwellian slang.

† http://www.thedailybeast.com/articles/2016/09/22/palmer-luckey-the-facebook-billionaire -secretly-funding-trump-s-meme-machine.html.

‡ Why did anyone spend serious cash subsidizing an activity that was already—as will be shown— incentivized? Mayhem was also instigated with rather small amounts of cash, as chronicled in the other examples cited here. This is an example of how even insiders are still finding their way. The online world has become so murky that no one has a complete view. The strange new truth is that almost no one has privacy and yet no one knows what's going on.

Do we really want to privatize the gatekeeping of our own public space for speech? Even if we do, do we want to do so irrevocably? Who knows who will be running Facebook when the founder is gone? Do billions of users really have the ability to coordinate a move off a service like that in protest? If not, what leverage is there? Are we choosing a new kind of government by another name, but one that represents us less?

The Invisible Hand on a Multitouch Screen

There is an even deeper question to consider. The attempts by the tech companies to battle shitposting comprise a fascinating confrontation between the new order of algorithms and the old order of financial incentives.

Old and new have a lot in common. The most enthralled proponents perceive each not merely as technologies invented by people, but as super-human living things. In the case of financial incentives, the elevation occurred in the eighteenth century, when Adam Smith celebrated the "invisible hand." In the case of algorithms, something similar occurred in the late 1950s, when the term "artificial intelligence" was coined.

The prevalence of shitposting and other degradations is fueled by the invisible hand, while the antidote is to be divined by artificial intelligence. So we will witness a new kind of professional wrestling match between the old made-up god and the new one.

Let's first examine how the old demigod, the one with the hand, influences behavior in the online world.

The business model that sustains companies like Google, Facebook, and Twitter is called advertising, but it's really something different. The model relies not so much on persuasion as on the micromanagement of human attention.

Such companies attempt to become the filter between a person and the world. This activity might sound like advertising, which is what it is called, but it really isn't. It is instead the only business model left once everything has become weightless.

Unlike advertising, social media and search's current business model is based not so much on biasing which *persuasive* information is most available; instead it depends on biasing the *options for action* that are most readily available, such as posts to read or links to follow.

The reason this works so profoundly well is the "cost of choice." The tech companies make their money by manipulating your perception of infinity. It would take an infinite amount of time to read and understand the agreement you click on in order to use the services, for instance, so you just click without reading.

Similarly, you can't look through millions of search results, so you accept that the artificial intelligence algorithms are the only option for charting infinite waters. The cost of choice, or rather the perceived cost of choice, becomes infinite when choices seem infinite. That is why so-called advertising customers pay so much money to companies like Facebook and Google. They rescue you from an infinite expense, but that means you let them partially make decisions for you. It mostly isn't about persuasion, but about biasing behavior in a more direct way.

The same design has been applied to the news.

A significant proportion of the population now receives news from a feed on a social networking service. You could keep many social media accounts, each with a different persona that would provoke a differently biased feed. But no one has the time, and besides, that would violate the policies of the contract you clicked on. So you must trust an intelligent algorithm, sifting through an endless ocean of news sources to bring each person the very best, most relevant news.

But the New Economy is in the process of gutting investigative journalism. The tech companies have won most of the cash that used to flow to newspapers for ads and subscriptions. Therefore, there are few genuine, high-integrity primary news sources, compared to antebellum days. There is almost no remaining local investigative reporting. There are occasional bloggers who accomplish real investigative work, but mostly they can only comment.

Steve Bannon claimed that "if *The New York Times* didn't exist, CNN and MSNBC would be a test pattern. The Huffington Post and everything else is predicated on *The New York Times* . . . That was our opening."* He couldn't have said this before the rise of the New Economy. The investigative press, which is distinct from the commenting class, used to be large and diverse.

---

* http://www.hollywoodreporter.com/news/steve-bannon-trump-tower-interview-trumps
  -strategist-plots-new-political-movement-948747

But a preponderance of people have fallen for a Netflix-like illusion* and believe that the main problem is too many news sources to sort through.

If there is little remaining investigative reporting, where does all that news, the apparently infinite supply, come from? It is brought to us by the invisible hand, aka old-fashioned financial incentives.

The world of viral posts, tweets, and memes is inherently detached from reality. They are catchy like pop tunes. No one fact-checks a pop tune. But that isn't the point. The point is that the devices on which this material is delivered track who is reading or watching at a given moment, and that's the truth that matters, not what's on the screen. Shitposting is more attached to reality than any previous form of communication, but the conveyance of reality flows from reader to server, not from server to reader.

The content that draws eyeballs is often alluring, even when allure isn't the main business, so the situation is confusing. We have to understand changes to user behavior as the product and the content as raw material for that product.

The apparent content of the online world—cute cats, baby announcements, unreliable news—is not the product. That stuff comprises the raw materials. And I'm by no means saying it's all bad. Love the cats, love the way people with shared issues can find each other. There's a lot of great stuff in the raw material.

But the product is different; it's the constraining of the easiest available options, to bias a user to buy something, do something, or believe something.

The Macedonian kids who made up nasty, false stories about Hillary Clinton made some money because they were selling raw materials, not yet turned into a product.† They drove traffic. The companies you ended up buying shoes or coffee from were the ones who paid tech companies, which acted as a gatekeeper, because your attention was gated. Your guided purchase was the product.

Once again, I am not saying there is no positive value in social media. Maybe the social media companies add enough value to justify their fees,

---

* Described earlier, in the section called "Birth of a Religion." Netflix uses AI recommendations to create the illusion that the selection of things to watch is larger than it is.

† https://www.buzzfeed.com/craigsilverman/how-macedonia-became-a-global-hub-for-pro-trump-misinfo?utm_term=.ghOlzDWAQ#.jj3XrKoY0

but whether they do or not, the point is that their income depends not just on getting the attention of a user for a little while, like old-fashioned newspapers, but for much of the time.*

With a real news source like the *New York Times*, I read, I get the news, and then I'm done. If the *Times*'s business model includes getting me to look at ads along the way, and perhaps be persuaded, great. But if the business model is to hold on to me to manage my choices for hours and hours of the day, then real news isn't of much use. It gets read—used up—too quickly.

Unlike the news, a news feed needs to get me cranky, insecure, scared, or angry. That's what will keep me in a Skinner box, where a service can manage which button is the easiest for me to reach.

The current business model of social media requires that it become part of the life of a user during all waking hours, even in the middle of the night if one can't sleep. Real news and considered opinions don't serve that goal well.† The sober contemplation of reality doesn't take up enough time.

Instead, a social media company must hold on to people by getting them angry, insecure, or scared. Or the service can place itself in between users and their friends and family, perhaps make them feel guilty. The most effective situation is one where users get into weird spirals of mob-like agreement or disagreement with other users. That never ends, which is the point.

The companies neither plan nor implement any of these patterns of use. Instead, third parties are incentivized to do the dirty work. Like Macedonian youth looking for some extra cash by posting poisonous fake news. Or even Americans looking for an extra buck.‡

Tech companies certainly never ask users to become thin-skinned, cranky, paranoid, or delusional; the snowflake persona just happens to be part of the emergent solution to the purely mathematical puzzle that *has* been explicitly posed: How do you drive the most traffic in order to take up the most time and attention?

It's worth noting that when social media isn't treated as weightless, but

---

* There are demographic differences to acknowledge. Older Americans apparently spend a great deal of time watching television, so in that case TV might serve more as a gate than a persuader. My argument is focused on younger generations that spend more time on cloud-connected gadgets.

† This sentence was written before the election of Donald Trump.

‡ https://www.washingtonpost.com/news/the-intersect/wp/2016/11/17/facebook-fake-news-writer-i-think-donald-trump-is-in-the-white-house-because-of-me/

as a source of professional value by its users, then there is less shitposting, as is the case on LinkedIn. Weightlessness is easy and fun, but a little gravity seems to bring out at least a few of the better angels in the natures of users.

To complete a sketch of how the two demigods interact, it must be pointed out that the weightless business model of social media is only one example of a trend in which companies use big computers to run exchanges in which they separate risk from reward. For another example, the bundlers of stinky mortgage-backed securities leading up to the Great Recession didn't want to know what they were selling, just as Google didn't want to know what its search page said about who won the popular vote in the USA's 2016 election. (The top story for a while after the elections was the false news that Trump had received more votes than Hillary Clinton.) To know is to take on liability, while to *not* know is to run a casino where everyone else takes the risks.

The Absurdity of Demanding That AI Fix Itself

What if the new demigod cannot knock out the old demigod? Maybe social media companies need to change how they make money. Maybe anything short of that is just a hopeless propping up of algorithms that will always be toppled by tides of financial incentives.

Just to be clear, I don't think ethical filtering can work, given the current level of our scientific understanding. Such fixes will just be gamed and turned into more manipulation, nonsense, and corruption. If the way to protect people from AI is more AI, like supposed algorithms with ethics, then that amounts to saying that nothing will be done, because the very idea is the heart of nonsense. It's a fantasy of a fantasy.

There's no scientific description of an idea in a brain at this time. Maybe someday, but not yet; so there's no way to even frame what it would be like to embed ethics into an algorithm. All algorithms can do now is compound what natural people do as measured by our impressive global spying regime over the Internet. And we're making a lot of those natural people into assholes.

But, just for the sake of argument, suppose the tech companies' attempts to fix shitposting with so-called artificial intelligence turn out to be spectacularly successful. Suppose that the shitpost-filtering algorithms are so excellent that everyone comes to trust them. Even then, the underlying economic incentives would remain the same.

The likely result would be that the next best way to drive cranky traffic would come to the fore, but the overall result would be similar.

As an example of a next-best source of crankiness, consider how Russian intelligence services have been identified by U.S. intelligence as meddlers in the U.S. election. The method was not just to shitpost, but to "weaponize" WikiLeaks to selectively distribute information that harmed only one candidate.

Suppose the tech companies implement ethical filters to block malicious selective leaking. Next up might be the subconscious generation of paranoia toward someone or something, in order to lock in attention.

If the companies implement filters to prevent *that*, there will always be some other method-in-waiting. How much control of our society do we want to demand from algorithms? Where does it end? Remember, well before we urged the tech companies to do something about fake news, we had demanded they do something about hate speech and organized harassment. The companies started booting certain users, but did society become any more temperate as a result?

At some point, *even if moral automation can be implemented*, it might still be necessary to appeal to the old demigod of economic incentives. There *are* alternatives to the current economics of social media. For instance, as I've suggested, people could get paid for their content over Facebook and pay for content from others, and Facebook could take a cut. (We know that *might* work because something like it was tried in experiments like Second Life, as described earlier.)

There are undoubtedly other potential solutions to consider as well. I advocate an empirical approach. We should be brave about trying out solutions, such as paying people for their data, but also brave about accepting results, even if they are disappointing.

We must not give up.

The Humane Use of Human Systems

It's not that the invisible hand is always going to be more beneficial than the imaginary AI being, or vice versa. Rather, people have to stop expecting perfection from *any* of the available demigods, which I'll define more explicitly as multiperson organizational systems.

Our Information Age favors the kind of person who's good at thinking like a computer, not that the mind-set is new to the world. It's the same mind-set that has always loved to commit totally to a system, to run a program; to seek the socialist paradise, the absolute theocracy, or the purist libertarian floating island colony where no one pays taxes and yet the island doesn't sink. If you can think like a computer program, you can win fortunes through the computer programs that now run the world.

To survive, mankind must also favor people who thrive in ambiguity and haven't pledged a blood oath to a single social organizing principle. Religions, markets, politics, cloud algorithms, societies, law, group identities, nations, education; we need them all and yet none of them are perfect. All these systems have failure modes, as an engineer would put it. The only way we're going to make it as a species is to think of our systems in the same way we think about our cars or refrigerators. Even the most reliable ones—even our cloud algorithms!—let us down once in a while. Purists have trouble accepting that.

If commerce might someday balance algorithms better, as I have suggested, there might be other checks and balances between our systems that can be articulated and explored. Just because algorithms are the most recent of the big systems, it doesn't mean that the others have suddenly ceased to matter.

There's a constant stream of pundits declaring that AI will destroy all the other systems in three minutes, so there's nothing left to do but enjoy a moment of feeling superior if you "get it" about our impending doom better than a normal, less technical person. But that is an abrogation of your responsibility, especially if you're an engineer.

We have wonderful systems to work with. If we don't dismiss the value left to us by every preceding generation, we can build a dignified, sustainable, very high-tech society that becomes the launching pad for adventures we can't yet imagine, but we engineers might have to learn a little humility to get there.

I've been careful in these final pages not to rely on assumptions about whether AI could ever get "real," meaning deeply, generally intelligent; or even conscious, for those who believe, as I do, that consciousness is real. The above arguments work out about the same whether AI will continue to require big data from people for the foreseeable future, or if it might start to function with a small amount of data, suggesting that it will have become more "free-standing."

For instance, the suggestion that micropayments could provide a sustainable, creative, and dignified alternative to the Basic Income Model will still make sense if AI becomes less dependent on big data.

Even if there comes to be general agreement that AI should be "doing all the work" in the future, the AI algorithms would still have to gather data from people in order to serve those people, unless humanity becomes utterly dull and predictable, or has effectively decided to commit mass suicide. So even if you are one of those nerdy souls waiting for a computer to write an ideal book for you, without having to mash up the words of human writers; even then, you could be paid for the data gathered from you that optimizes that algorithmic writing for you, and in that transaction there would be a distribution of power and wealth in the society and at least a small fountain of dignity.

I know, I know, the counterargument is that the computers will just kill the people. When I hear this common retort, I am transported back to the old Little Hunan, where I urged my friends to think about the guinea pig in the flamethrowing tank. You can interpret the Internet as having come alive already. You can interpret the election of a confused president in the USA as a method the Internet used to start clearing out the people. There isn't a supernatural ontology police force that will swoop out of the sky to scold you. But interpreting humans as having responsibility is the only one of the available interpretations that gives us a chance to take responsibility.

Don't make the mistake of treating this book as a conservative or traditionalist reaction against trendy futurism. I'll usually out-futurize other futurists in futurism cutting contests.

My futurism is real, while a lot of the stuff out there is fake. A futurism that suggests a total break with the past—a singularity, or an AI taking over—is a fake. A break with the past just means starting over, making ourselves primitive. Indeed, we've already shown that we can make ourselves primitive online. Maybe we can make ourselves crude enough that an algorithm seems superintelligent in comparison.

The McLuhan ramp proposed at the end of appendix 1 is only one example of a future vision that is at least as colorful as AI supremacy fantasies, but genuinely looks to the future, instead of a retreat to the past.

A futurism that pretends we already know science that we don't know is also fake. When someone pretends we already know everything important about how brains work, that's being a fake futurist. What's really being said is that the person is committed to being stuck forever in the ideas of the present.

I remain dismayed when someone assumes AI is already on an inevitable course that will lead to it needing only small amounts of data—instead of vast mountains of stealthily stolen data—in the near future.

If you ever come to a lecture where I am addressing students, you might hear me remind them of an example brought up earlier in this book; that at the end of the nineteenth century, there were confident proclamations that physics was *done*. Then came the twentieth century's general relativity and quantum field theory, and those theories are incompatible with each other, so we know physics is *still* not done. I tell students that science is the thing that kicks your butt. I tell them that it's impossible to be a scientist without being able to acknowledge the unknown.

Our fate rests on human traits that haven't yet been defined in scientific terms, such as common sense, kindness, rational thought, and creativity. While the AI fantasy is that we'll be able to automate wisdom any minute, can we all at least agree that these qualities can for now only be *harnessed* by our systems? That they can't be *generated* as yet by our systems?

The questions of our age: Can we see through our seductive information systems in order to see ourselves and our world honestly? How bad do things have to get before tech culture decides it's worth challenging even our most cherished mythologies in order to dig ourselves out of our mess?

Acknowledgments

Some passages of the present work are adapted from my contributions to John Brockman's edge.org or my anthologies of science writing. Other passages are adapted from my works originally published in the *Whole Earth Review* or the *New York Times*.

My wife, Lena, not only supported and put up with me while I wrote this book, but I am astonished at the strength and brilliance she displayed during a period when she was battling cancer. Thank you!

Thanks to Maureen Dowd for correspondence that inspired some passages.

Thanks to Satya Nadella, Peter Lee, Harry Shum, and everyone else at Microsoft Research for their camaraderie and support. Not a word here represents a Microsoft point of view, of course.

Thanks to Mary and Steve Swig for the writing cottage at The Shadows.

Thanks to my editors, Gillian Blake in the USA and Will Hammond in the UK, for their always graceful interventions, and especially for their patience during a year of delays and difficult circumstances; and to my agents, especially Jay Mandel. Thanks to Eleanor Embry at Holt for her tireless attention to manuscript details.

Thanks for reading early drafts: Michael Angiulo, Tom Annau, Jeremy Bailenson, Steven Barclay, Maureen Dowd, George Dyson, Dave Eggers, Mar Gonzales Franco, Edward Frenkel, Alex Gibney, Ken Goldberg, Joseph Gordon-Levitt, Lena Lanier, Matthew McCauley, Chris Milk, Jane Rosenthal, Lee Smolin, Mary Swig, and Glen Weyl.

Illustration Acknowledgments

All images in this book are courtesy of the author with the exception of the following:

Page xiii   Photographs by Kevin Kelly, used with permission.
Page 2   © AP Photo / Jeff Reinking.
Page 44   Top: © Mark Richards. Courtesy of the Computer History Museum. Bottom: Courtesy of the Inamori Foundation.
Page 99   Courtesy of Steve Bryson.
Page 120   Photographs by Ann Lasko Harvill, used with permission.
Page 121   Left: Photograph by Ann Lasko Harvill, used with permission. Right: Photograph by Kevin Kelly, used with permission.
Page 125   © Linda Jacobson.
Page 128   Photograph by Walter Greenleaf, used with permission.
Page 130   TK
Page 131   Reproduced with permission. Copyright © 1987 *Scientific American*, a division of Nature America, Inc. All rights reserved.
Page 132   Photograph by Dan Winters. Courtesy of *Scientific American*.
Page 133   Courtesy of Wikimedia Commons.
Page 140   © REX / Shutterstock.
Page 166   Drawings by Ann Lasko Harvill, photographed by Kevin Kelly, used with permission.
Page 177   © George MacKerron, used with permission.
Page 181   Reproduced with permission. Copyright © 1984 *Scientific American*, a division of Nature America, Inc. All rights reserved.
Page 190   © MixPix / Alamy Stock Photo.
Page 193   Photograph by Ann Lasko Harvill, used with permission.
Page 194   Photograph by Kevin Kelly, used with permission.
Page 214   © NASA.
Page 215   Photograph by Young Harvill, used with permission.
Page 218   © AP Photo / Eric Risberg.

Page 222   © AP Photo / Oinuma.

Page 231   Photograph by Kevin Kelly, used with permission.

Page 270   Top photographs: © Rick English Pictures. Bottom: Photograph by Ann Lasko Harvill, used with permission.

Page 272   © Douglas Kirkland / Getty Images.

Index

Page numbers in *italics* refer to illustrations.